Spatial Sense Makes Math Sense

Spatial Sense Makes Math Sense

How Parents Can Help Their Children Learn Both

Catheryne Draper

ROWMAN & LITTLEFIELD
Lanham • Boulder • New York • London

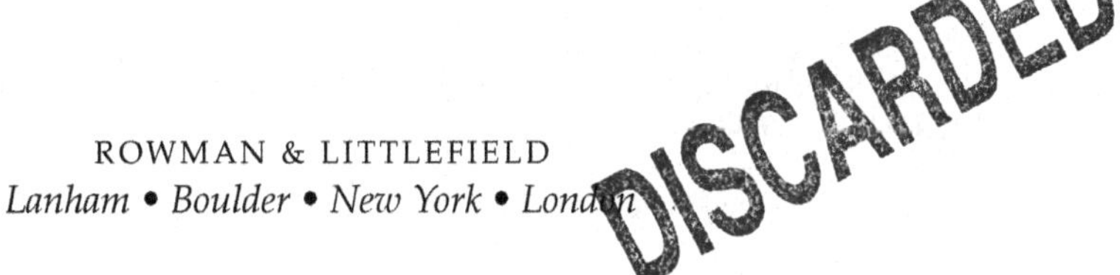

Published by Rowman & Littlefield
A wholly owned subsidiary of The Rowman & Littlefield Publishing Group, Inc.
4501 Forbes Boulevard, Suite 200, Lanham, Maryland 20706
www.rowman.com

Unit A, Whitacre Mews, 26-34 Stannary Street, London SE11 4AB

British Library Cataloguing in Publication Information Available

Library of Congress Cataloging-in-Publication Data

Names: Draper, Catheryne, author.
Title: Spatial sense makes math sense : how parents can help their children
    learn both / Catheryne Draper.
Description: Lanham : Rowman & Littlefield, a wholly owned subsidiary of The
    Rowman & Littlefield Publishing Group, Inc., [2018] | Includes
    bibliographical references.
Identifiers: LCCN 2017033640 (print) | LCCN 2017047262 (ebook) | ISBN
    9781475834307 (Electronic) | ISBN 9781475834284 (cloth : alk. paper) |
    ISBN 9781475834291 (pbk. : alk. paper)
Subjects: LCSH: Mathematics--Study and teaching (Elementary) |
    Education--Parent participation. | Spatial ability in children.
Classification: LCC QA135.6 (ebook) | LCC QA135.6 .D7297 2018 (print) | DDC
    372.7--dc23
LC record available at https://lccn.loc.gov/2017033640

♾️™ The paper used in this publication meets the minimum requirements of American National Standard for Information Sciences—Permanence of Paper for Printed Library Materials, ANSI/NISO Z39.48-1992.

Printed in the United States of America

Eric
Rachel
George
Jacob
Brendan
Jack
Joey
Kristen
Adam
Andrew
RICHARD
MADELINE
Talia
Ted
Danielle
Kelly
Cody
Connor
Nate
Andy
Caitlin
Russell
Rick
Jessica
Bradley
Doug
Ken
Katie
Ryan

Connie
Katie
CARLOS
Ethan
Robbie
Emma
Shane
Jaclyn
Joshua
Sydney
Jane
Nell
Ren
Joan
Kathleen
ADRIENNE
Mary Beth
Don
Liv
Whitney
Max
Matthew
Geoffrey
Chandra
Timothy
Elizabeth
Connor
Angela
Betty

Mary
Jared
Roger
Hannah
Jonathan
Neil
Walker
Tom
WILL
Amber
Becky
STACEY
Mitch
Ben
Rob
ADA
HARRY
JESSICA
Erik
Teresa
Maggie
CHARLES
KATY
ANNA
Paul
Brendon
Marie
John
Joanni

Jena
Maggie
Kaylee
Lexi
Olivier
Madeline
Grace
Kajsa
Bianca
Melissa
Jason
JENNIFER
Wendy
Richard
EJ
Joy
Alex
Cathy
Roberta
JAMES
Kevin
Margaret
Deirdre
Megan
Nancy
Ricky
Peter
Rachel
Mallory

Elizabeth
Nathan
Heather
Mora
Christine
Bobby
Francis
Nicole
Jay
Janis
Jacqueline
Jason
Rachel
Teddy
Randy
Eda
Dominique
Mark
Catie
Doug
Derek
Lauren
Joe
Lisa
Dana
Diane
Roseanne
HEIDI
LORY

Mandy
Sean
Jo
Deborah
Ed
Kendra
Brian
Seamus
Chris
David
Peter
Ralph
Carlo
David
Raymond
Phillip
Chris
Jeffrey
Dan
Shawn
Brandon
Tanun
Sean
Steven
C. J.
Jean
MATHEW
Taylor
Jeff

Evan
Lindsey
Reiss
Libby
Kim
Nicole
Dee
Colleen
Lisa
Traci
WILL
Holly
Leigh Ann
John
Beth Ann
Stephani
Kelly
DEBORAH
Chloe
Mallory
Terri
William
Devan
Jan
Lee
Shane
Robert
CAMILLE
Molly

Dedicated to all the students who taught me how to teach them.

Contents

List of Figures

Foreword

"When I grow up, I think I will be a teacher. I will probably be teaching arithmetic because it is my favorite subject."

This is the two-sentence, prize winning essay I wrote as a first grader for my hometown newspaper (the prize was a $1.00 check). I keep the essay and check stub in a frame in my office to remind me that while the study of mathematics typically begins with learning the arithmetic of counting, number facts, and number fluency, it goes far beyond "just" working with numbers. As a six year old, I certainly didn't realize that I was on a journey to develop my math sense that included number sense and so much more…including Spatial Sense.

Catheryne Draper's book, *Spatial Sense Makes Math Sense,* will be especially useful for all parents, whether Math Avoiders or Math Aficionados (to use Cathy's descriptors), who want to help their children use their visual strengths and develop their spatial sense. If you've always thought of math learning as primarily the memorization of a set of steps to use to perform computations at each grade level, then this book will open your eyes to what you can do (and enjoy doing) to help your children build conceptual understanding of mathematics across their school career that will enable their success in all grades from early elementary through high school and calculus.

Until this book was written, I had not found any published material to explain to parents the importance of their child's development of spatial sense, the same spatial sense that prepares them to succeed in their school work and perhaps even pursue a career in a STEM (Science, Technology, Engineering, or Math) field that relies heavily on spatial sense and visual approaches to problem solving.

In this book, Catheryne Draper presents explanations of many geometric concepts across the grades and offers suggestions to parents that will help them work with their children to develop their visual and spatial senses. The time that parent and child spend doing math at home, in the park or on a walk will help them build their math understanding, especially of geometry concepts and vocabulary to become strong math learners who will begin to make connections between classroom geometry and geometry in the world around them.

As I read the first chapter and the following chapters, I found myself talking out loud to Cathy as if she and I were together in the same room discussing the ideas and suggestions she wrote about in each one. I especially related to her quick anecdotes about specific students she'd taught and/or tutored over the years to illustrate how and why she made the suggestions she did to address these students' needs. I read, highlighted and took notes as a math professional, middle school and high school teacher, and even more importantly as a parent and grandparent who wanted to share and build the concepts and math understandings with my family.

Every chapter spoke to me and I feel that they will speak to parents, too. For example, in Chapter 3 (Sorting and Classifying the Symbols of Space) Cathy discusses the vocabulary of geometry, which often challenges our students, especially students who enter school as English as second language learners. She offers guidance in ways that parents can support their children in seeing the math around them and in learning the shapes and solids vocabulary in the real world and in the classroom. I related strongly to this chapter based on my work as a math coordinator in my school district.

I was fortunate to visit many classrooms over the past two decades. I observed students with their teachers in grades K through 12 making connections between the 2 and 3-dimensional geometry they see in nature, in pictures and in man-made structures around them. On a beautiful New England day one recent autumn, I joined a group of first graders and their teacher walking briskly in the schoolyard and around the neighborhood. They were walking and talking, using geometry vocabulary appropriate for early elementary students to describe what they were seeing. This is certainly something a family could do to develop a child's vocabulary and connect it to the names and properties of shapes.

Each of the four books written by Catheryne Draper in this well-planned and well-written series is designed to meet the needs of parents in supporting their own and their children's math abilities and understandings. We can all thank Cathy Draper for writing *Spatial Sense Makes Math Sense: How Parents Can Help Their Children Learn*. It will definitely guide parents as they work to reduce or eliminate their and their child's math anxiety while developing their math understandings. Our students' success in math class in school and beyond is a goal we (parents, teachers, and the author) all share. This book can help us in meeting our goal.

Carol Hynes

Preface

The sadness in the mother's voice as she told me she never had the chance to experience any of the fun that her child was now having in math is one of my main reasons for writing the four books in this series. This parent told me that she had always been stuck in the drudgery of remedial math classes. She admitted that had she known some of the things that her son was experiencing now, then her own math life might have been very different. Her despondency, and others like hers, served as a powerful motivation to try to change parents' negative emotions about math. Lowest common denominators need not be gatekeepers to enjoying mathematics.

Parents *can have the opportunity to enjoy* the interesting relationships and patterns in mathematics and share them with their children. "Fortunately, you [and your child] can understand and enjoy mathematics without having to find square roots, or memorize the quadratic formula, or prove geometric facts."[1] The "good stuff" seems to be reserved for the ones who are "good" at math *already* because they can reel off the "math facts" at Mach speed even on cloudy Mach-changing days, add any fractions regardless of the denominators, or annihilate any and all long-division problems.

Facts, fractions, and long-division trials and tribulations should not be the barriers to the "good stuff" or the "right stuff" of mathematics for you in your past or for your child in the present. As the previous three books in this series have emphasized, imagination is the best and possibly the most important, if not *the most precious*, learning resource, and imagination should be encouraged at every possible opportunity.

The three previous books in this series (*Winning the Math Homework Challenge: Insights for Parents to See Math Differently, User-Friendly Math for Parents: Learning and Understanding the Language of Numbers Is Key*, and *How the Math Is Done: Why Parents Don't Need to Worry About New vs. Old Math*) all set the pace for *Spatial Sense Makes Math Sense: How Parents Can Help Their Children Learn Both* by using the same four Big Ideas of Definition, Organization, Relationships and Patterns, and Connections as themes for understanding the ins and outs and connecting threads across many familiar, and some not-so-familiar, elements of the "good stuff" in mathematics.

This book describes mathematics through students' stories, all of them true, real-life experiences; but the names were changed in order, as one student said, that they wouldn't be found out. The stories describe misconceptions, confusions, and misunderstandings as well as some delightful insights, descriptions, and downright courageous efforts. Many are visual-spatial learners, some are auditory-sequential learners, and all of them either shared a common frustration of confusion, or a need to explore more and deeper levels of mathematics.

As the Math Avoider, children started to give themselves permission to move spatial figures and symbols around, as did the children in the series' earlier books, and they also started to take risks and even conjecture an insight or two. The children began to feel like they may indeed *have some control* over their math learning! As their understanding improved, so did their math voice, their thinking, and in many cases, the scores on those repeatedly recurring tests.

Parents only want the best for their children, and many want their children to have a better time in math than they did themselves. Like the parent mentioned at the beginning of this preface, many adults wish they could have had fun too. My hope is that these books provide that opportunity for some fun, for you *and* your child. I also hope that this book gives you permission to use your spatial strengths, ask questions *that don't have immediate answers*, and join the mathematicians past and present who welcome the learning opportunities that come from making mistakes, solving problems in a new and different way, and making your own math discoveries.

C. D.

NOTE

1. Jacobs, *Mathematics: A Human Endeavor*, xii.

Acknowledgments

It takes a village to complete a project as broad reaching as a four-book series, especially a series about mathematics designed for Math Avoiding parents currently faced with helping their children. Our village was filled with incredibly insightful readers, helpful librarians, kind and careful critical advisors, vigilant editors, patient artists, and encouraging friends, and of course students and their parents. I am indebted to the many people who carefully read through the manuscripts, especially the Math Avoider readers who struggled through their own past math issues to make sense of these current explanations. Some read the manuscripts multiple times just so the descriptions and the images could have the clearest possible explanations and figures.

These books would not have been possible without the talent and support of Iris Berry, Katie Finch, Carol Findell, Robyn Frost, Ann Garabedian, Mary Ann Grassia, Carol Hynes, Nell Hartley, Tom Hartley, Eric Knight, Kim Knight, Pat and David Lindholm, Dana Linn, Rita Losee, Johnny Lott, Ann Hammond McCamy, Kathy Miles, Marcia Perry, Regine Philippeaux Pierre, Sally Russell, Robin Schwartz, Les Warrington, Rudy Weekes, and Mary Weisenberger, and with special thanks to Linda Bouchard from Snow Harbor Graphics and Kate Victory Hannisian from Blue Pencil Consulting. I also want to thank all of the students who shared their feelings, confusions, understandings, some "Aha" moments, and, in a few cases, Herculean efforts to climb over their math tribulations.

To all of those who rekindled my spirits when the creative fires dimmed, I cannot thank you enough! To all who cheered me on to achieve those deadlines, I am so very grateful. To all of you who freely and generously shared your ideas and your knowledge in your particular fields, including your encouraging and kind cautions to keep explanations clear, accurate, and uncomplicated, I have a deep appreciation. I hope our efforts worked. You be the judges.

Introduction

Were you better at algebra or geometry, or were both of them an enigma? Very often parents and adults who are out of school will proclaim that they were either geometry people or algebra people—they understood one while the other was always unfathomable. On occasion, you will hear someone declare that both were confusing until calculus, when everything came together. *Spatial Sense Makes Math Sense: How Parents Can Help Their Children Learn Both* brings the strengths of both algebra (arithmetic) and geometry into focus by showing how the algebra can be explained with the spatial relationships in geometry.

And still there are others who never recognized a purpose for any of these topics or, in truth, in any mathematics, at least until now when they are facing math again as they try to explain the homework to their children. To help parents with this task, *Spatial Sense Makes Math Sense: How Parents Can Help Their Children Learn Both* shows how developing spatial sense can help *visually explain* all of those math relationships. You will read about Sophie Germain, who believed that algebra and geometry worked hand-in-hand because, as she described them, algebra is a written form of geometry and geometry is algebra with shapes.

Some prefer to express their algebra as pictures like Germain, some prefer just symbolic reasoning, others look forward to the logic in the systematic argument and proof that math requires and Math Aficionados relish. The last chapter in the last section provides a way for former Math Avoiders (if you were one earlier, then hopefully by the time you reach that section you will have at least started to peek around the Math Aficionado door) to "see" how calculus can wrap up those topics learned in earlier grades into a reasonable arrangement that can bring these earlier topics together to make sense.

This book takes you through similar stages as the ones traveled with numbers in *User-Friendly Math for Parents: Learning and Understanding the Language of Numbers Is Key* and *How the Math Gets Done: Why Parents Don't Need to Worry About New vs. Old Math*. Beginning with the Big Idea about interrelationships between shape sense and symbolism, this book continues with an exploration of the Big Idea about dimensions as an organizer of shapes. The third section of the book focuses on shape relationships and patterns, and Part IV closes out the Big Idea theme with connections among and between math concepts.

In *User-Friendly Math for Parents*, we made a distinction between number symbols and number sense, so it is only fitting that we make a similar distinction between shape symbols (shape pictures) and shape sense (also known as spatial sense). Being comfortable with shape arrangements, transformations, and other geometric combinations helps children make sense of their surrounding space and translate that to a geometric system.

The three chapters in the Definition section (Part I) for *Spatial Sense Makes Math Sense* introduce spatial sense and some of the related geometry vocabulary. Chapter 1 explains a perspective of spatial sense and how this awareness of space affects learning, especially for the visual learner. Chapter 2 addresses the sometimes-problematic vocabulary issue and explains how some of the geometry words came into being. Chapter 3 chunks the large amount of geometry vocabulary into manageable classification categories.

The Organization section (Part II) takes a trip into the nineteenth-century satire called *Flatland*. The journey begins in Chapter 4 with the shape inhabitants in Lineland and describes why these inhabitants are measured with one-dimensional units. Chapter 5 travels to Flatland with two-dimensional shapes and explains how two-dimensional shapes evolved from Lineland's one-dimensional shapes to generate areas in Flatland. Chapter 6 crosses the border from two dimensions into three dimensions (from Flatland into Spaceland) and describes several versions for volume.

Part III, the Relationships and Patterns section, brings all three dimensions into the relationship-and-pattern conversation. Chapter 7 describes the significance of two categories of shapes, "similar" and "congruent." Chapter 8 introduces some visual explanations for pi (π), phi (ϕ), a limit idea (e), and eccentricity (ε). Chapter 9 explains more shape patterns, some even surprisingly predictable.

The final Connections section (Part IV) shows how three topics significantly connect early elementary math to more sophisticated math ideas. Chapter 10 describes how that early-known, right-triangle version for pyramids continued to influence mathematics and measurements from ancient Egypt through the Pythagoreans days in Greece and on to an expansion in mathematics called trigonometry. Chapter 11 shows how the seemingly simple circle has a powerful reach into just about anything that involves angle degrees. Chapter 12 brings arithmetic and algebra together again to remind us that math recycles topics to build on prior knowledge.

All four books in this series encourage flexibility, and this book carries the flexibility theme into spatial sense with shapes so that you and your child can experience how your imaginations can play a significant role in understanding geometry. Mathematicians and Math Aficionados are still innovating and creating and adding to geometric ideas. While we don't know what tomorrow's geometry will look like, we do know that spatial visualization is an essential asset for tomorrow's geometry to evolve. Maybe even your budding Math Aficionado will be the next Lobachevsky or Riemann to develop another non-Euclidian geometry.

Part I

DEFINITION

Geometry—the science of furniture and walls.[1]

—W. W. Sawyer

Geometry is the most abundantly available learning resource in our dimensional world's existence because everything has a shape, just like furniture and walls. Geometry is the branch of mathematics that deals with shapes in space, both by measuring them and by identifying relationships among them. Some shapes have auspicious attributes assigned to them, like Plato's dodecahedron, a shape that you will meet in Chapter 1 and then revisit in Chapter 6, as the "guiding scheme for the harmony of the Cosmos."[2] Even if the harmony of the Cosmos is not one of your central concerns, the harmony of your child's experience with geometry homework certainly is.

Number symbols represent the ideas behind number sense, so it should follow that shape symbols (shape pictures) represent the ideas behind shape sense (also known as spatial sense). Numbers are the symbols related to calculations just as shapes are the symbols related to geometry. Having a sense of composing and decomposing works with both number sense and spatial sense. Spatial sense helps to understand how shapes are "artfully" composed and decomposed. This first section describes how to decipher those symbols in order to use them to communicate in mathematics.

Chapter 1 provides a perspective on spatial sense, what it is, how it shows up, and why it is helpful in learning mathematics in general, but especially in geometry. Chapter 2 takes on the task of learning geometry vocabulary by addressing the derivations and other vocabulary-type "ancestry." Chapter 3 describes two organizational strategies that help to "chunk down" the large amount of new vocabulary. So many terms to learn in so little time; a good reason for your child to start using their classifying skills to their advantage.

The terms for the spatial symbols are used throughout the chapters, with brief context descriptions for their introductory meanings. More complete definitions of the terms are referenced when they are explained in the book. All of the terms used across all geometry topics cannot be incorporated into just one book, but enough terms are provided to give you and your child a firm foundation on which to build further knowledge and more sophisticated vocabulary and symbols. Euclid formally established the structure for definitions that your child and classroom curricula still use today.

Euclid provided the math community with thirteen books called *Elements* that included a structure for geometry not previously organized in this way. His definitions and propositions (sometimes referred to as axioms or just another way to say that a statement is true) are still used in today's classrooms. His books outlined definitions and propositions that described geometric structures and terms that mathematics and mathematicians have relied on for centuries. It is interesting that two thousand years after Euclid, along comes Oliver Byrne, who reexplained Euclid's *Elements* using pictures—a visual-spatial influence indeed!

Friedrich Fröebel also incorporated visual strengths by developing nineteenth-century "Kindergarten Gifts"[3] to help early learners learn about and use the language of space. Froebel's Gifts have been remanufactured under a variety of names and brands, yet all still provide the same basic explorations "to develop visualization skills through hands-on experiences with a variety of geometric objects."[4] From nineteenth-century Froebel to twenty-first-century math-teaching standards, spatial sense has been essential to math development. Anything you can do to encourage imagination and curiosity is worthwhile, especially in geometry.

NOTES

1. Sawyer, *Mathematician's Delight*, 10.
2. Ghyka, *The Geometry of Art and Life*, 114.
3. Brosterman, *Inventing Kindergarten*, 40.
4. NCTM, *Principles and Standards for School Mathematics*, 43.

Chapter 1

A Perspective on Spatial Sense

There was the week in Mrs. Reine's fourth-grade class. I was allowed to help construct a model colonial village. I made houses, a stockade, and an intricate little spinning wheel. It was the most glorious week in elementary school.[1]

—John Dixon

This story actually began back when the author, the now Dr. John Dixon, was finishing first grade and threatened with the stigma of not being promoted to second grade, a fate too often experienced by many children who are spatial learners. The math that they encounter is often just about symbols that are meaningless until these children have a grasp (literally) on the *spatial arrangement of numbers and the associated numeric* meaning of the symbols. The models of the houses and the intricacies of the spinning wheel pieces made sense because they *spatially* fit together. Spatial sense, simply speaking, is when things make sense in space.

Does everyone have spatial abilities? Yes, to some degree. Are some children more gifted in spatial abilities than others? Yes; we all have our special gifts. Howard Gardener launched his theory of intelligences over two decades ago, spatial intelligence being one of his original seven intelligences. Can you develop your child's spatial abilities? Certainly. Is it necessary for your child to be spatially intelligent or spatially gifted in order to understand mathematics? Probably not, but it certainly helps to understand the interconnectedness of shapes and other things that can be represented "spatially" so that they "fit together to make sense."

"Spatial-visualization . . . [is one] of the most often noted interests of future mathematicians."[2] While your child may not be mathematician-bound, spatial-visualization and spatial sense will certainly make the math that your child is doing *now* more interesting. Geometry is filled with shapes having math relationships, an especially rich environment for expanding your child's spatial abilities. This chapter is all about building upon those existing spatial abilities using geometry's concepts, and the spatial relationships that they invite in math. Spatial abilities need to be encouraged in any way possible, as early as possible for as long as possible.

As a homework-encouraging parent, you can provide opportunities to invite imagination to help your child build spatial thinking processes to make relationships make sense, like the fitting together the intricate pieces of spinning wheels. By using both imagination (mental) and building (physical) activities, your child will learn how to *visualize* concepts with spatial sense, *think about* shapes and concepts in different ways, be comfortable with flexible ways to *see and express* what they are seeing, and have that *glorious experience before fourth grade*. Earlier exposures to spatial-oriented activities might have averted young Dixon's first-grade trauma.

Using spatial-driven strategies for understanding concepts is essential for visual-spatial learners; these kinds of instructional activities are also valuable for auditory-sequential learners. Both kinds of learners deserve to have the benefits of understanding concepts and expressing the concepts with math symbols. Without understanding, the marks for numbers or shapes are simply squiggles on a page, whether a student is just beginning the learning journey in first grade or venturing into another world with new symbols in calculus. Seeing is the beginning of expressing, and seeing from different perspectives, promotes flexibility of expression.

In order to learn more about how young children saw and described spatial images, Grayson Wheatley provided several spatial assessments in elementary classrooms.[3] He showed children some images for a brief three-second period, and then he removed the image from their sight and asked the children to describe what they had seen. The images and the children's descriptions are shown in Figure 1.1. These responses could be anyone's, not just those of young children. It should not be a surprise that your child, and others, see the same images but with different and more personal versions of their own sense of space around them, or spatial sense.

Math Avoiders will be happy to know that spatial sense does not have right or wrong ways to see figures. However, the flexibility with which your child *describes these objects or shapes* can significantly improve the learning experience; the more, and different, perspectives, the better. Your child will be at a distinct advantage in geometry, and in the rest of mathematics, if they can imagine a given figure (or shape or object) from different perspectives or orientations, and then generate appropriate,

"I saw a hexagon with a big X in it."

"I saw two diamonds and two triangles."

"It looks like an envelope opened up."

"It looks sorta like a cube but a line is missing.

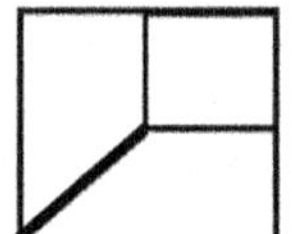

"It looks like a room."

"I see a hallway."

"It looks like a pyramid with the top cut off."

"I see a small square connected to a big square."

"I see a skylight."

Figure 1.1   Children's comments about figures.

possibly even different, descriptions for the figure. By the way, do you see one large rhombus (diamond) in Figure 1.1 after the "envelope" flaps are opened?

Spatial sense has several associated labels: spatial visualization, spatial reasoning, spatial perception, and visual imagery, to name a few. Spatial sense is also about the interpretations of what you see. Both girls and boys have this basic, hardwired spatial ability and both will benefit from building with blocks or mechanical toys, arranging pictures and designs with colored blocks, talking about the same shape in different sizes, sequencing blocks or shapes by size, and categorizing shapes by specific properties (properties can be thought of as adjectives, or the adjectives can be thought of as early-onset properties).

One of the goals of helping children develop spatial sense, or improving what they already have, is to help them recognize the interconnections of something in their space as part of *their* "interconnected system"[4] around themselves. Interconnectedness is recognizing how objects and concepts are intertwined in relation to your own "world." What children see and describe *from their perspective* plays an integral role in their adjusting to their spatial world around them.

Playing one of those round-robin visual and verbal games such as "I Spy" will help your child piece together spatial descriptions *in their mind* while developing a vocabulary around spatial awareness. Also, you will learn a great deal about *what* your child is seeing and, most importantly, you will learn that *the way your child sees their world will probably not be the same way that you see it.* Remember, spatial awareness and spatial sense is not "right or wrong," but it can be developed, nurtured, and improved.

One example of how a father created an early-development spatial vocabulary activity with his child was by helping his two-year-old son Sean classify his toys. Sean had developed his own version of the language of space by learning how to describe and compare the things that he saw in his world. Sean's dad made a math activity out of helping Sean clean up his toys by telling Sean to put all the trucks (using Sean's definition of what made something a truck) in one container and all the balls (by Sean's definition of what constituted a ball) in another container.

Sean had already developed a clear sense in his mind of the difference between a truck and a ball, so the classifications were clear *in Sean's mind*. Sean's dad helped Sean understand two concepts with just one activity, the life skill of cleaning up *and* that ever-so-important classification skill for math! As your child gets older, the descriptions have more refined details, but the importance of classification and spatial vocabulary stays the same. As the category details get more sophisticated, more containers are needed to sort more items with more specific categories. By doing this, your child is learning how to look for geometric properties.

A creative kindergarten teacher designed an activity for helping children visualize and record natural spatial events. The teacher was trying to give her children a visual-spatial sense for time. Her class marked the sun's location on the window using tape. At regular intervals during a day, one child would place tape on the window where the sun shone through. If window taping is not possible, then marking the sun's shadows on a nearby wall at different intervals will also give an image of time change as the sun moves across the sky. Time on a round clock is a visual arrangement, too, but could be a bit premature for very young children.

PERMISSION TO MOVE FREELY

Using flexibility with a sense about spatial arrangements helped a high-school student named Theo give himself permission to see diagrams in different ways and orientations. Theo was stymied by the drawing of a straight angle (straight angles have 180°) marked into three angles with one of the angles marked (with the small box) as a right angle (right angles have 90°). He was trying to solve a test question that asked for the measure of *the sum* of the other two angles given the image in Figure 1.2. The total measure of those other two angles was an enigma until he saw the rearrangement of the angles in a right triangle. No explanation was necessary.

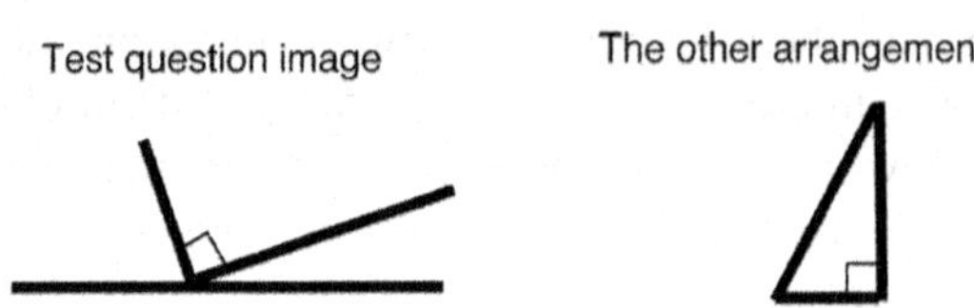

Figure 1.2   Theo's test question rearrangement solution.

Later, Theo admitted that he did not know that rearranging the three angles was a "legal math move"—he didn't have "permission" to think flexibly. Knowing which angle in the triangle matches with which of the other two angles on the straight line is not necessary to answer the question, but recognizing that all three of the angles must total 180° for either the triangle or the straight angle *is* necessary. All Theo needed was to recognize that the other two angles must equal 90° in order to complete the 180° total. Simply redrawing it made the problem make sense. Chapters 2, 3, and 11 include more explanations and relationships about angles.

A seventh-grader named Lizzie was learning about right angles in triangles. She could "say" that a right angle measured 90° and she had measured angles on paper with a protractor, but when asked to *make* a right angle with strips that snap together like AngLegs (straws, toothpicks, or Exploragons would work just as well), Lizzie did not know how to use a protractor to identify the 90° measure in this different situation. After Lizzie *used a protractor to help her make a right angle*, she found the correct length to finish making a right triangle. Unfortunately, Lizzie was also under the misimpression that it wasn't a right triangle anymore when *this same triangle was rotated to a different position*.

Both Lizzie and Theo are typical of students who have not had the opportunity to learn about the benefits of being flexible in their sense of space thinking. For whatever reason, the math in their respective worlds had been rigid and without "legal moves" to move items and maintain equal measurements in different orientations. Flexibility in thinking about shapes can be nurtured by showing that all shapes are assembled or composed from other shapes. Shapes don't just appear out of nowhere. Starting with a triangle, the first shape in two dimensions, you and your child can assemble several more shapes.

COMPOSITIONS AND DECOMPOSITIONS OF SHAPES

"Every polygon can be built from triangles, and most other shapes . . . can be approximated by polygons."[5] This quote by Ian Stewart, a noted twenty-first-century mathematician,

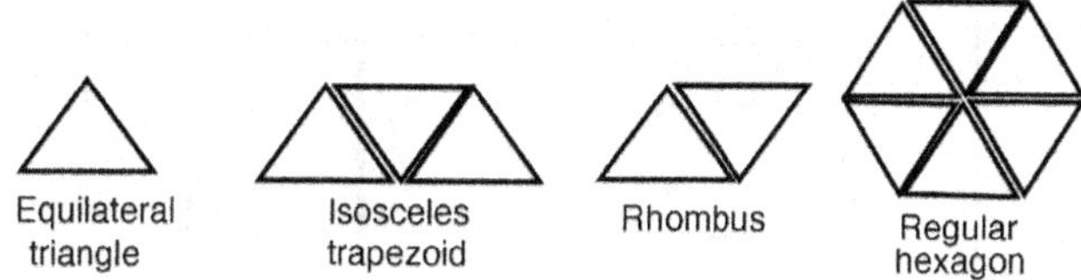

Figure 1.3   Equilateral triangle image compositions.

would invite your child to try to generate some convex (no angle in the shape is greater than 180°) shapes with triangles. Start with easy shapes first. It makes more sense to start composing shapes with the least number of sides that can make a convex two-dimensional "regular" shape, such as the equilateral triangle shown in Figure 1.3. Shapes that have all equal side measures and all equal angle measures are called "regular."

The equality term used with measurement in geometric shapes is replaced with the term "congruent" and congruent refers to the *shape and size comparison* between two shapes. For example, two triangles (trapezoids, pentagons, or any polygon or polyhedra) are congruent if all sides and angles of the two triangles (pieces of the polygon or polyhedra) exactly match each other. Having congruent sides, angles, and diagonals (faces, edges, and vertices of polyhedra that you will read about in Chapter 9) in a shape controls the shape name and how the shape is used in other shape relationships. Congruency is about shapes; equal is about numbers.

Since equilateral triangles have specific measures, then they can make (or take apart) only specific shapes that you will read more about in Chapters 7 and 8. As you build these shapes, remember that these shapes can just as easily be taken apart using the same triangles (or different ones). Building and taking apart at the same time helps to build the relationship between the two ideas. Assembling shapes seem to be easier than taking apart, especially if the shape in later years just shows the outline. More descriptions of the shapes are explained in Chapter 2; for now, the idea is to show how geometric shapes are composed and decomposed.

After making shapes with an equilateral triangle, you and your child can try to compose shapes with other kinds of triangles and look for comparisons. A right triangle is the next-most familiar type of triangle, so use it to build a rectangle, a parallelogram, an isosceles triangle, a right angle trapezoid, and even another isosceles trapezoid like

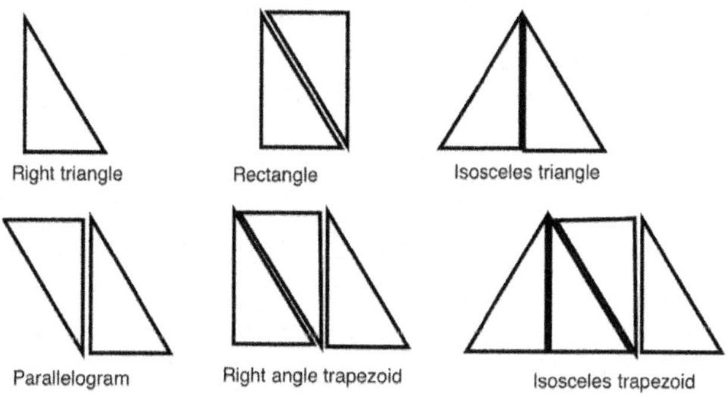

Figure 1.4   Right triangle image compositions.

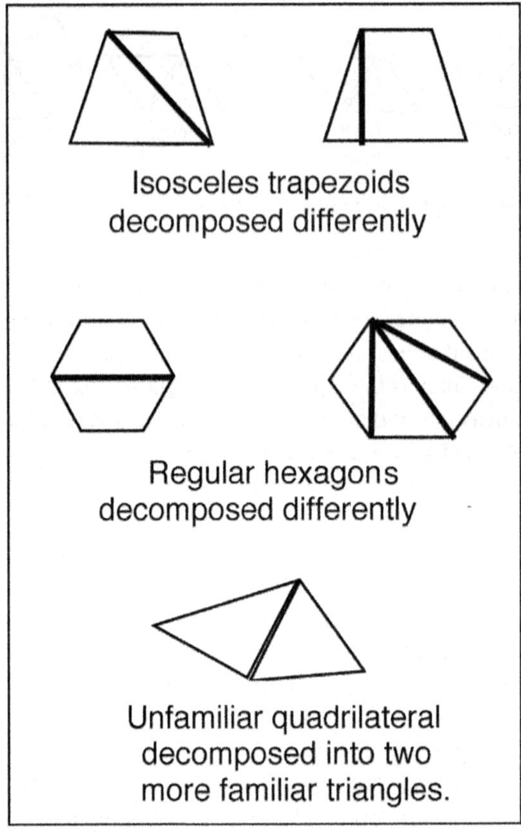

Figure 1.5   Different decompositions.

the ones in Figure 1.4. The regular triangle, like other regular polygons, must have all sides and angles congruent, but the only requirement for a right triangle is having one 90° angle. These geometric shapes and related vocabulary in these examples, and other shapes, are explained with more descriptions in Chapters 2 and 3.

After composing these shapes, your child can also *decompose these same shapes back into equilateral triangles, isosceles triangles, or other shapes*, depending on what may be needed, or easiest to see, to solve a given problem. Different problem questions will necessitate different arrangements at different grade levels. For example, you can decompose that regular hexagon into four triangles or into two isosceles trapezoids. Or you can decompose an isosceles trapezoid into a right triangle and a right-angled trapezoid, or into two unlikely triangles as shown in Figure 1.5. You get the idea. The only limitation is in your child's imagination!

Whatever decomposition will help your child answer the question in a given problem is the one to use. As for using the math terms like "trapezoid" or "rhombus" (instead of "diamond"), remember that your child can probably say "Tyrannosaurus Rex" and "Apatosaurus" easier than you can and from a very early age. Compositions and decompositions are much easier to "see" if your child is given "permission," like Theo and Lizzie, to arrange the shapes into different orientations and *still keep the same measurements.*

> Ask Yourself:
>
> Could you make even more convex shapes using equilateral triangles?
>
> What would they look like?
>
> What would shapes look like if you used isosceles triangles that were not equilateral?

Figure 1.6   Task Box: Ask Yourself.

These familiar triangle shapes can also be mixed and matched to make other more unusual shapes. Having this flexibility in composing and decomposing shapes will make many formula-driven calculations, and even solving word problems that refer to shapes, a great deal easier to remember and solve. Your child, and even Math Avoiders, may not have a problem building shapes or even naming the shapes. However, taking these same shapes apart seems to present a different kind of difficulty.

The shape names may seem a bit quirky perhaps, but nevertheless reasonable because the name of the shapes are closely related to the number of sides of the shapes. You will find the explanations for these names in Chapters 2 and 3. Math Avoiders tend to think that there is only one "right" way to decompose shapes, the same way they may have been lulled into the notion that there is only one "right" answer or one "correct" way to solve a problem. As Stewart reminds us, there are many ways to use polygons to build other polygons.

> Try This:
>
> Build a rhombus with several copies of the same size right triangle. Give yourself permission to rotate the right triangles to build your rhombus.
>
> Try to build a regular hexagon with lots of right triangles. Experiment with different right triangles.

Figure 1.7   Task Box: Try This.

SYMMETRY IS SPATIAL, NOT ONE-HALF

Some helpful practice books and a few textbooks try to make the symmetry concept easier with a short sometimes-true description, such as "symmetry is ½." Symmetry is a reflection *in space*; one-half is a number, a ratio *between numbers*. Symmetry describes a shape duplicate or reflection like the leaded window on the left in Figure 1.8; one-half is a number ratio used to *compare one quantity* with *two equal quantities totaled*, like the half examples in Figure 1.8 that *are not all symmetric.* Saying symmetry is one-half

Symmetric
Window   One-half

Figure 1.8   Symmetry vs. one-half.

might make sense to an adult, but it can send the wrong message to children, especially while they are learning those early numbers.

When a line of symmetry is drawn on the shape and the two images on either side of that line reflect across that line, much like you would see if you folded the shape in the middle (vertical fold for the window), then the two sides of the shape do have the same area measure. In that case, one of the two area measures could represent one-half of the total area. Several more symmetric examples in the hexagon graphic organizer (Figure 1.9) show how symmetry is a description, not a number. The topics in the organizer indicate how shapes *and other topics* can have different symmetries, some with one line and others with two or more lines of symmetry.

In fact, reflection symmetry is a description of an *image reflection with respect to a line*, whether the line is drawn *on the image* or when the entire shape or image is reflected over an external line, like you might see in a mirror. In each case, the line is reasonably called "a line of symmetry." The external lines of symmetry, and other symmetry transformations such as rotations and slides, are described in Chapters 7 and 8.

A vertical line of symmetry in a *normal curve* graph goes through the mean (average) middle number in the graph. The positive and negative directional *distances* between the integers on a number line repeat so the line of symmetry would be drawn through zero to represent symmetry. (If the numbers continue infinitely then the total would have an $\aleph$ symbol, then ½$\aleph$ would be another infinity!) Symmetry doesn't require adding numbers or infinity, much to everyone's relief.

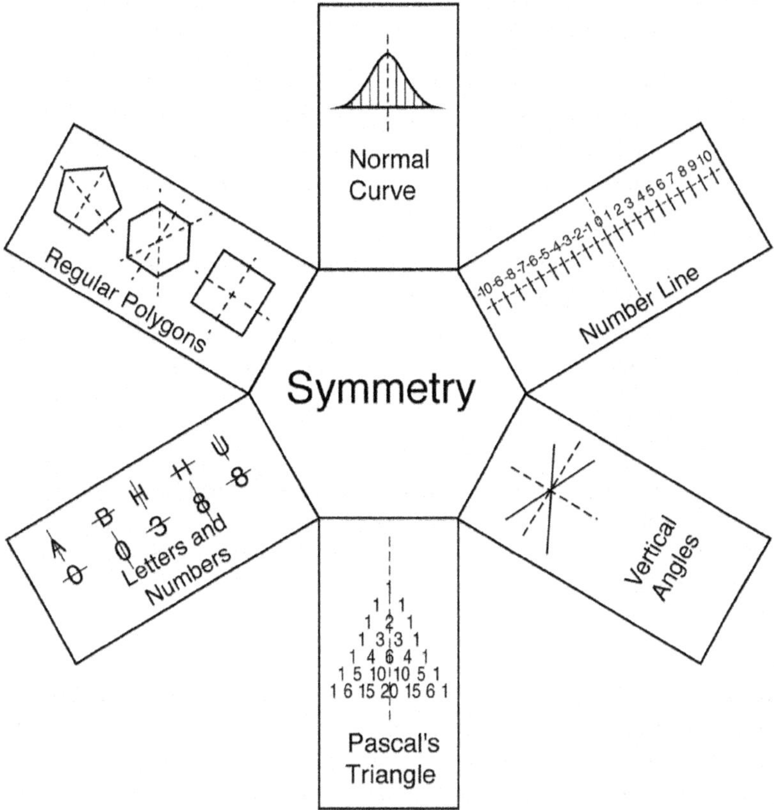

Figure 1.9 Hexagon web diagram for symmetry.

Vertical angles (possibly named vertical rather than horizontal because the two crossing lines that make them look like a "v") can have two lines of symmetry. The dotted lines in Figure 1.9 show how the vertical angles "shape" can be folded vertically *and* horizontally so that the fold creases show how the two sides match exactly. Having a good spatial sense (or sense of space or a sense of how things are arranged in space) allows your child to recognize vertical angle symmetry, and other nonpolygon shape symmetry, in different orientations or within other shape arrangements. No numbers to add, just mental ways to fold.

Pascal's triangle has a vertical line of symmetry *of the number locations and the triangle shape.* As a history note: just as the Pythagorean theorem was a known relationship before Pythagoras proved it, the triangle array of numbers in Pascal's triangle was known earlier than when Blaise Pascal's name was attached to it. A Chinese mathematician wrote about this interesting arrangement and relationship among numbers in the fourteenth century. About four hundred years later, Pascal developed many more connections in mathematics for this triangle of numbers, both algebraic and geometric, thus earning the memorialization with his name.

Some of the English-language letters and digits have more than one line of symmetry, suggesting a connection between type font and early Arabic numerical notations.

> Prove It to Yourself:
>
> Get a piece of tracing paper or patty paper or parchment paper used for baking. Patty paper, sometimes called tracing squares in the store, got its name from the thin sheet that goes between hamburger patties.
>
> Choose a regular polygon and trace it on a piece of patty paper. Fold the traced shape so that sides and corners match both sides. Continue folding to find all of the lines of symmetry for your regular polygon.
>
> Trace and fold an isosceles rhombus, a parallelogram, or a non-square rectangle.
>
> How many lines of symmetry (fold creases) do these shapes have?

Figure 1.10   Task Box: Prove It to Yourself.

The letters and number symbols, like regular polygons, can be folded to match sides so that the folds themselves show how the shapes duplicate on either side of a line of symmetry. More information about these multiple lines of symmetry for regular polygons appears in Chapter 7 when the symmetry topic is associated with congruence.

KEEP IN MIND

Numbers are related to arithmetic as shapes are related to geometry; numbers are the symbols used in arithmetic operations and shapes are the symbols in geometry. Having a spatial sense of how shapes are combined (*and* numbers in number sense) allows composition and decomposition (taken apart and put back together again) and gives children one extra step ahead as they artfully maneuver through spatial awareness—and toward more sophisticated mathematical and geometrical relationships.

NOTES

1. Dixon, *The Spatial Child*, 6.
2. Hersh and John-Steiner, *Loving + Hating Mathematics*, 21.
3. Wheatley, "Image Maker: Developing Spatial Sense," 375.
4. Dixon, *The Spatial Child*, 9.
5. Stewart, *Taming the Infinite*, 86.

Chapter 2

The Vocabulary Clues Are in the Words

There is hardly a culture, however primitive, which does not exhibit some rudimentary kind of mathematics.[1]

—Davis and Hersh

There you have it—no escape from math in our cultured world. Language is necessary for communication in this world, so learning mathematics also means learning how to "speak math." Some suggest that math has its own language with specific adjectives, nouns, and verbs. Geometry, then, as a spatial version of mathematics, could be said to have its own spatial language. This chapter is about learning the clues to this new language, a language that almost demands imagination to "see" the shapes.

Understanding how geometry vocabulary might have developed and had a continuity of understanding over the time span of different cultures would have required a common, or at least consistent, language that could transcend cultures and time. Our current math terminology has the Greek and Roman cultures to thank; however, the earlier Egyptian, Babylonians, and Chinese certainly contributed their influence. Imagine the kind of difficulty that could happen in communicating ideas across the miles reaching from Greece to Egypt, and across from Sicily to Turkey without twenty-first-century technology of Internet, email, and social media.

Words, terms, and descriptions *had to carry their message within the word itself.* Too much was at stake to risk any ambiguity. The development of civilizations depended upon accuracy and interpretation. The duration of time and length of distance required clarity, especially when it took centuries to develop ideas. How we currently use geometry words and terminology to describe shapes, relationships, and orientations in space is the focus of this chapter. What may sound like a totally different math language, in actuality, is not so different from Latin and Greek roots.

The words (combined with their related adjectives) in geometry are the clues to unlocking much of the seeming mystery in geometry's vocabulary that your child faces today. This chapter provides a kind of etymology (could be referred to as vocabulary ancestry) for many of the seemingly strange-sounding mathematical terms that your child must learn. Understanding the "ancestry" of the prefixes and the words will also help your child have a basis for deciphering words and terminology for future math studies.

The shape names include a prefix that identifies the number of sides, and then those shape names can also tag on an "adjective of arrangement or measurement" to describe a relationship of the measures of their angles and sides or both. Other mathematical words and terms also have descriptions included as part of their arrangement orientation or measures. Knowing how to interpret the construction of the names and terms opens those comprehension doors to the mathematical language.

PREFIX POWER FOR POLYGONS AND POLYHEDRA

The prefix is the clue that identifies the number of sides of polygons or faces of the polyhedra shapes and the names that we use have a Greek or Roman origin. They probably used the same prefixes because the two-dimensional polygons ("poly" for many and "gon" for corner or angle) make the faces of those three-dimensional polyhedra (same "poly" and "hedra" for many-sided). The web diagram of names in Figure 2.1 shows how the numbers along with the Greek and Latin prefixes influence the names.

If your Latin or Greek isn't that great, then capitalize on English-language words that also have similar number-driven prefixes, such as tricycle for three and octopus for eight. The familiar month names of September, October, November, and December have roots in these prefixes as do several other, some less-familiar, words in the English language, such as tripod, triangulation, triple, quadrangle, pentachord, pentacle, pentameter, hexachord, septenaries, decimal, enneagram, and the Dodecanese Islands, just in case you were wondering.

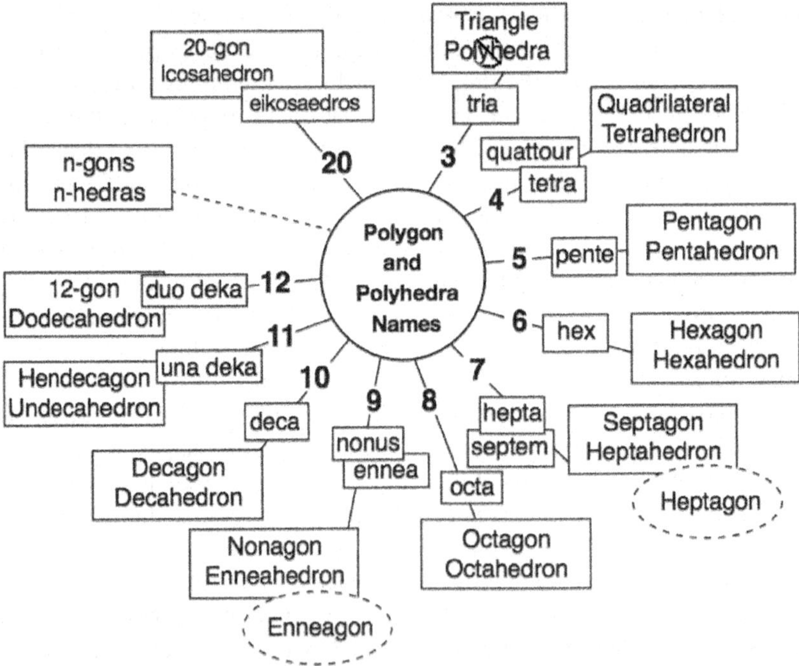

Figure 2.1   Polygon and polyhedra name web diagram.

A reasonable question to ask is why the twelve-sided shape was assigned a Greek or Latin prefix name while its adjacent neighbor with thirteen sides was relegated to the number prefix as a thirteen-gon label. It could be that those shapes with more than twelve sides were assigned an "n-gon" name (when the "n" is the number of sides) just for convenience. Choices had to be made about the use of prefixes for names. If "quadtour" for naming a two-dimensional shape as quadrilateral was already taken, then they may have preferred a different prefix that indicated four sides to assign to the three-dimensional polyhedron with four faces, so they chose to name it a tetrahedron.

You can blame the Greeks for not providing more Greek-language prefixes for polygon names or the Romans for not naming polygons or polyhedra with Latin prefixes, but on the other hand, be pleased to not have to remember more names! That "n-gon" number prefix or "n-hedron" does the trick, is clear, and identifies the number of sides of the polygon or polyhedra. When Plato started exploring with congruent faces and angles in the same solid, he jumped to twenty faces and named the icosahedron, leaving the ones in between with just number names. Other mathematicians have played with solids, too, and their solids are described in Chapter 9.

Your child will learn about spheres and cylinders as three-dimensional curved solid figures; however, these solids are not considered polyhedra because *polyhedra are limited to straight-sided faces*. While the rounded sphere may be relegated to just the category of three-dimensional solid, it enjoys a relationship with pi that polyhedra do not have. The same straight-sided limitation applies to two-dimensional circles; a circle is not considered a polygon because it does not have straight sides.

A three-faced polyhedron is not on the polyhedra list either, and neither is a two-sided shape on the polygon list. That is because four is the minimum number of faces in three-dimensional space and three sides is the minimum number of lines for a two-dimensional shape. More explorations and explanations about how the number of dimensions influences the kinds and names of shapes are described in Chapter 3.

Many more polygons and polyhedrons are just waiting for your child to explore. Sculptors and other artists have found them fascinating, as have architects, engineers, scientists, and mathematicians. The names of polygons and polyhedra may share common prefix names, but there is a significant difference between polygons and polyhedra with their associated number of angles. The three-dimensional polyhedrons do not share the same number of corners (also called vertices, one vertex per corner) as their faces (see Figure 9.8 in Chapter 9), but two-dimensional polygons do have the same number of sides as angles. More details are in Chapter 9.

VOCABULARY ANCESTRY

As the homework-helping parent or grandparent, you are faced with an onslaught of new vocabulary information even after you get beyond just naming the shapes. In fact, naming the shapes is probably the easiest part. The magnitude of the task of learning all those new words and all those new formulas is a frequent complaint about geometry and can be a daunting chore. If you or your child has this complaint, then you are probably exhausted from learning all that previous what-felt-like unrelated math information. Relief is here simply by connecting the derivations to the descriptions.

Ask Yourself:

Do all polygon sides *have* to have the same length?

Can you draw a polygon that has different side lengths? Use only convex polygons.

Do all polyhedra faces *have* to have the same shape?

Can you make a polyhedra with different faces?

Is there a maximum number for "n" for a polyhedron with "n" faces or polygon with "n" sides?

Answer:
Polygons can have different side lengths. Only "regular" polygons must have all sides equal. The same is true for polyhedra.

Figure 2.2   Task Box: Ask Yourself.

The Greek culture, and later Latin and then early French influence, gets a lot of credit for the descriptive nature of our math vocabulary "ancestry." Like those polygon and polyhedra names, much of the English-language vocabulary shares common prefixes and descriptive adjectives from the same sources. Two examples for geometry words that children frequently confuse are "parallel" and "perpendicular." "Parallel" stems from the Greek "para" for alongside combined with "allēlos" for one another. "Perpendicular" has a plumb line origin that refers a Latin reference to hanging, "perpendiculum."

Plumbers and carpenters have used plumb lines for millennia. Twenty-first-century technology may not use the string and plumb line but it still incorporates the concept of gravity making the line straight as it hangs vertically down with the weight just heavy enough to keep the string straight. Lures at the end of a fishing line also use these same types of weights so that the line drops vertically underwater. Our Roman friends incorporated their word into the geometry language; what is it about "perpendicular" spelling that could be your clue?

If the Greek and Latin origins are not to your liking, then possibly some fabricated mnemonic "ancestry" elements might help. For example, perimeter is the distance around the edge or rim of a shape, so look to the actual spelling of the word for a clue—"perimeter" contains "rim" in the spelling. "Parallel" spelling has two "ls" in the spelling that are reminders for parallel lines. Some of the creative ideas in the spelling can also be reminders for those Greek or Latin beginnings, too, like "aequilaterus" is the Latin origin for "equilateral" and "equiangular" and the equal reminder is also part of the spelling for the words.

Having a "word wall" for displaying vocabulary has been part of many classrooms of all grade levels. This idea also works well at home, either on the always-dependable refrigerator door or on an actual designated wall in a playing area. Children tend to be very creative about ways to remember words and their definitions when given

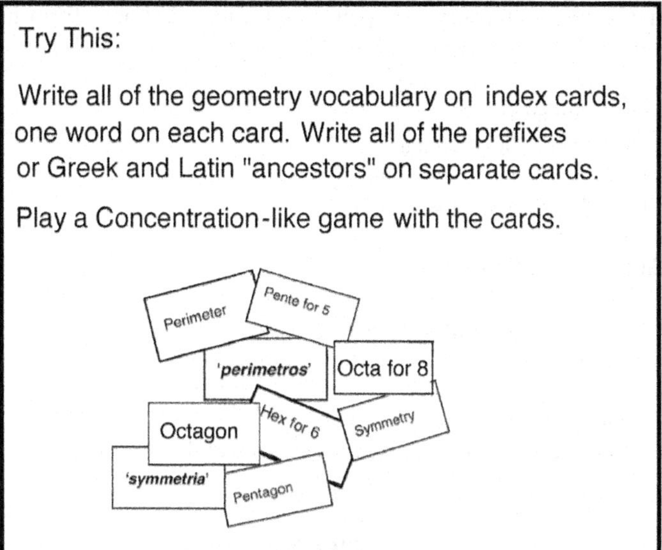

Figure 2.3   Task Box: Try This.

the chance. A valuable side-benefit for children being involved in understanding the words in geometry-speak is the questions that they ask during the experience. Jules asked about a rhombus, "Why can't a rhombus be a kite?" Jules was trying to make rhombus (to her, it was still a diamond) make sense. Can it be?

The definitions of geometry words cannot be separated from their "ancestry" or etymology of shape terminology. They must go hand-in-hand; however, making the effort to understand *the thinking behind* how these terms were selected can help with understanding the orientations, the arrangements, the measurements, and the relationships of space. Each of the chapters that follow contain many more geometric terms, and all of the terms are explained and organized with the idea of helping your child develop her own version of word "ancestry."

KEEP IN MIND

The geometry shapes and terms have survived over many cultures so they are not likely to go away any time soon. These words were created in such a way that would reflect the measure, the arrangement, or the relationship between sides and angles of the shapes, regardless of the cultural time line or the distance between cultural sites. In this twenty-first century, children are learning the language of geometry so that they can express their thinking about current geometric ideas and possibly think new thoughts about new and different geometric ideas. Appropriate language is integral to the communication of math ideas.

What better way to wrap up a chapter about what we know from history than to provide a few of the lines from a student's poem:

Egyptians and Babylonians
Gave us the names over 5000 years ago.
They developed a sophisticated understanding of geometry,
It is them we must thank for what we know.
On to modern geometry
We give credit to Ancient Greece.
It was Thales and Pythagoras who changed their thinking,
And Euclid who published the elements of what they did seek.
Euclid's book organized the Greeks' mathematical thoughts.
The Elements were their path to deductive reason.
Thirteen books and over 600 theorems,
These are what they were sure of, just like the changing of the season.

NOTE

1. Davis and Hersh, *The Mathematical Experience*, 9.

Chapter 3

Sorting and Classifying the Symbols of Space

Learning is easy, it's remembering that is difficult.

—Kermit the Frog

"Chunking it down" is an expression used to help reduce the overwhelming feeling associated with accomplishing an enormous task. Chunking is another way of expressing classifications into smaller categories, so maybe chunking and classifying can help manage the enormous task of learning so many new words for geometry. Besides, "memory is intelligence-specific,"[1] so using learning tasks that promote visual learning strategies is far preferable than rote memorization or "practice makes perfect" repetition, neither a guarantee for remembering for a spatial learner, or Kermit.

Organizing the vocabulary words into categories makes it a lot easier to learn, clarify, and remember. While the classifications chosen for this chapter are based on two-dimensional arrangements and measurements, the idea of classifying into categories also applies to three-dimensional terms. By organizing these new words and information, your child is using a valuable critical-thinking classification strategy, and is also laying down a memory trail for retrieving information.

These categories can help answer some of the "why" questions that happen during the beginning stages of learning new information and new vocabulary. Since geometry is figured algebra (see Germain's opinion in Chapter 12) and describes space, then it is reasonable to sort the words into categories that are associated with arrangements and measurements of geometric space. Arrangement-driven words eventually have measurements and the measurement-driven words begin with relationships among the shapes. After all, that is what the Math Aficionados like to do, play with measurements and relationships.

ARRANGEMENT-DRIVEN VOCABULARY

The terms in the web shown in Figure 3.1 show a sampling of words in an arrangement-driven category. These terms can serve as a starting point for how arrangements have an impact on the names for these shape terms, *and* help with vocabulary ancestry.

21

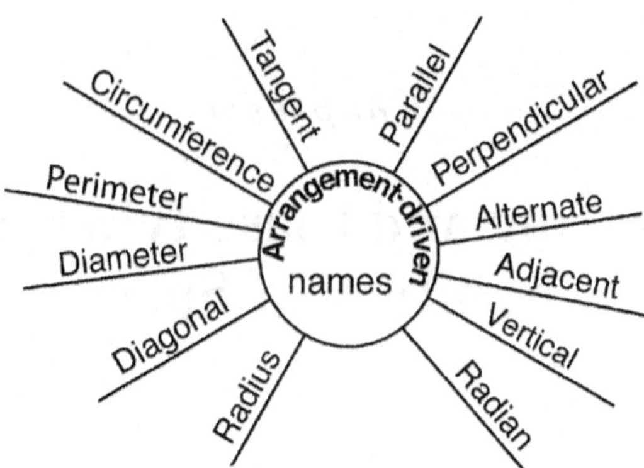

Figure 3.1   Arrangement-drive name web diagram.

Your child can *make up* categories when new vocabulary appears or when trying to make sense of current vocabulary. Classifying as an organizational strategy helps in learning new geometry words while at the same time improving your child's spatial descriptions. Arrangement-driven terms are described in later paragraphs, some were mentioned in Chapter 2, and others are described in later chapters.

Parallel and Perpendicular

These two words seem easy enough for those who are already familiar with what they look like. But for a visual child just learning what these two words mean, there can be a real issue in remembering which word goes with which arrangement. Visual images combined with imagination can encourage your child to generate his own interpretations and definitions for these two terms, just as fifth-grader Alicia did for herself when she made the collage posters shown in Figure 3.2. With camera in hand and an eagerness for searching, Alicia set out to find and collect examples of parallel and perpendicular from all around her environment.

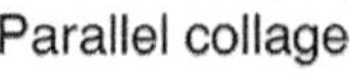

Parallel collage

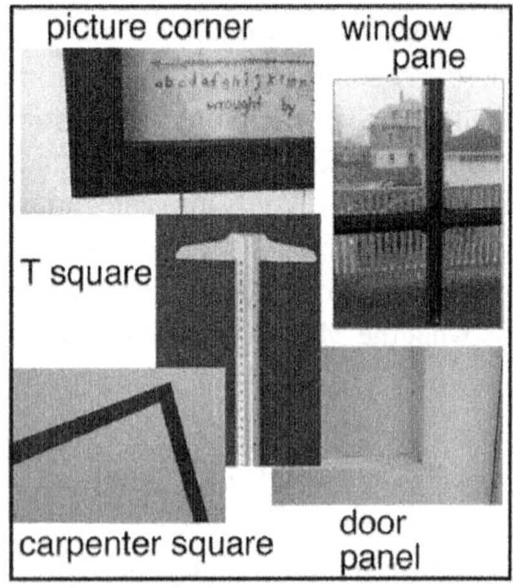

Perpendicular collage

Figure 3.2   Alicia's parallel and perpendicular collages.

It is interesting to note that the 90° angle measure that we cling to as an essential description of perpendicular was not part of Euclid's third-century BC landmark treatise of thirteen books called *The Elements.* In the first book, Euclid defined two lines as perpendicular "when the two adjacent angles [angles next to each other and made by the two lines] are equal to one another."[2] Euclid then follows that phrase with, "this arrangement makes the two adjacent angles right angles." Whether your child uses Euclid's adjacent angles or the 90° measure for perpendicular will likely depend on your child's age and grade level.

Remember that "parallel" gets into the shape name game because of the arrangement of the *line segment sides* of a shape, such as parallelogram. Other shapes have parallel sides but may not carry the name, so in those cases, the term becomes part of the shape's properties. Properties of a given shape are all of those "adjective" descriptors of that shape. For example, some shapes that have parallel sides as property descriptors are trapezoid (one pair), rectangle (two pairs), rhombus (two pairs), and square (two pairs). Be careful to pay attention to *how many pairs* of sides are parallel.

It gets even more interesting if the shape with parallel sides has adjacent perpendicular sides or even diagonals with a perpendicular arrangement. That's a mouthful of words for what is just an arrangement relationship! That is where the 90° angle measurement can come into play. While a rectangle can have perpendicular adjacent sides and two pairs of parallel sides, not all sides in all rectangles are necessarily congruent and the diagonals are not perpendicular. All sides congruent *and* opposite parallel sides belong to the square. Adjacent sides of that same square are perpendicular *and* that square has perpendicular diagonals. Interesting, indeed!

Alternate, Adjacent, and Vertical

These three words have the same math meaning that they have in the usual English language. These terms are most often associated with angles and sides describing how the angles, or sides, are arranged in reference to each other. Alternate interior angles, adjacent angles, and vertical angles all refer to angles. There are other arrangement-driven words, such as opposite, interior, exterior, and central to name a few. The reason to focus on just three of these is to provide a setting and a starting point for deciphering the others, and relate them to expressions you already use.

Alternate angles, like using alternate hands with the do-si-do square dancing, mean that the two angles are on either side of a line. The usual reference is to alternate interior angles of a transversal (line) of parallel lines when pairs of angles are located on either side of the transversal ("trans" indicating that the transversal will "cross over" both parallel lines). Alternate sides of the street is usually used to describe parking restrictions for street sweeping regulations. Its word ancestry, not surprisingly, also comes from Latin *alternus* "every other" or *alternare* "done by turns."

Adjacent angles, like adjacent sides, are angles (or sides) that are next to each other or adjoining each other. Adjacent (from Latin *adjacere* for "lying near") is used to describe locations concerning houses, such adjacent neighbors or adjacent rooms. Adjacent angles are next to each other and even share a common side (just as

Figure 3.3   Diagonals and Diameters.

neighbors share a common fence or hedge) and a common corner vertex. Adjacent angles appear in polygons and in complementary and supplementary angles (described later in Chapter 4). Adjacent sides appear in polygons, such as adjacent legs in a right triangle or congruent adjacent sides of a square. Math-speak, indeed!

"Vertical" is often used to describe two angles oriented in space, such as vertical angles. Vertical also defines a line's orientation in reference to a horizontal position. Vertical sides would likely refer to the sides of a three-dimensional container or possibly the sides of a two-dimensional shape or orientation of two-dimensional parallel lines. Parallel lines, or any single line, could be described as either vertical or horizontal when describing a shape or setting in a word problem.

Circumference, Diameter, Perimeter, Diagonal

Circles originally got named from Latin *circulus* for "small ring," and other English words continue the use with "circus" (All the circus events happen in the round center circle under the tent!) and circumnavigate (sail around the world). It isn't a strange phenomenon that the perimeter of a circle would be named "circumference" using the Latin *circumferentia* or *circum* for "around," and *ferre* for "carry, or bear" as a vocabulary "ancestor."

A diagonal does not exist without a polygon's perimeter and a diameter does not exist without a circle's circumference. Diagonal and diameter both represent going across the shape; the "dia" Greek prefix represents "going across." A diagonal goes across and connects two vertices of a polygon and a diameter goes across a circle through the center to the other edge of the circle (Figure 3.3). Both of these, diameter and diagonal, control their respective shapes, although some might say the shape controls the diagonal and diameter.

MEASUREMENT-DRIVEN VOCABULARY

An angle is created within a polygon with only two sides as line segments. Angles have their own measurement-driven vocabulary that describes the measures of the angles, such as obtuse, acute, and right angles. Your child can think about their arms as a way to make angles. If one arm is outstretched and one arm overhead (see Figure 3.4), then your child will have that right-angle position that the Babylonians referred to as "looking *right* up." As your child rotates that "right up" arm downward, then other angles are created.

Figure 3.4   Flag positions as angles.

The two arms of the stick-figure diagram make the angles in the upper right section, and the arms continue to make acute angles until the two arms meet. Just like other patterns, these angle measures repeat in all four sections around the body, like those used by ship flag-signal messages and the airplane signaling ground crew. The obtuse angles are made with the arms as the two arms begin to go down the left side of the stick figure until the two arms are in opposite directions. The two right "Babylonian position" angles reasonably create the 90°+90° or 180° straight angle (the "R" signal in the flag alphabet).

These quarter sections around the body are called quadrants when arranged in a coordinate graph or used for the rest of the flag alphabet. Acute angles are the angles that the arms make in the first quadrant (all less than 90°), and the obtuse angles are the angles that arms would make in the second quadrant (greater than 90° but less than 180°), as long as one arm stays in the straight-out right position shown in Figure 3.4. Those angles in the third and fourth quadrant are used in later math studies.

Then when these angles are part of specific triangles, the triangles take on their names as descriptions as shown in the web in Figure 3.5. The angle measures control the lengths of the opposite sides, and the sides of the triangles now start having a say in the naming of these triangles, such as scalene, equilateral, and isosceles. Quadrilaterals are also composed when four sides are assembled with angles, some take on the same descriptive definitions that the triangles have, such as isosceles trapezoid or right-angle trapezoid.

The right triangle gets special attention because it frequently appears in compositions of shapes due to its special relationships of sides, more fully described in Chapter 10. A right triangle appears when two sides (also known as legs) of that triangle meet at a 90° vertex. The third and longer side that is located across from the 90° angle is called the hypotenuse. When a small box is in the angle, the 90° is guaranteed. The Greeks named the hypotenuse from "hypo" (meaning under) and "teinein" (meaning stretch) and the origination of the term "legs" is credited to a Norse term "legger" probably because, like people, there are two of them.

MIXING ARRANGEMENTS AND MEASUREMENTS FOR NAMES

Let's talk now about how mixing the angle measures with side measures increases the vocabulary to describe shapes. Start easy with triangles. Because a triangle has only

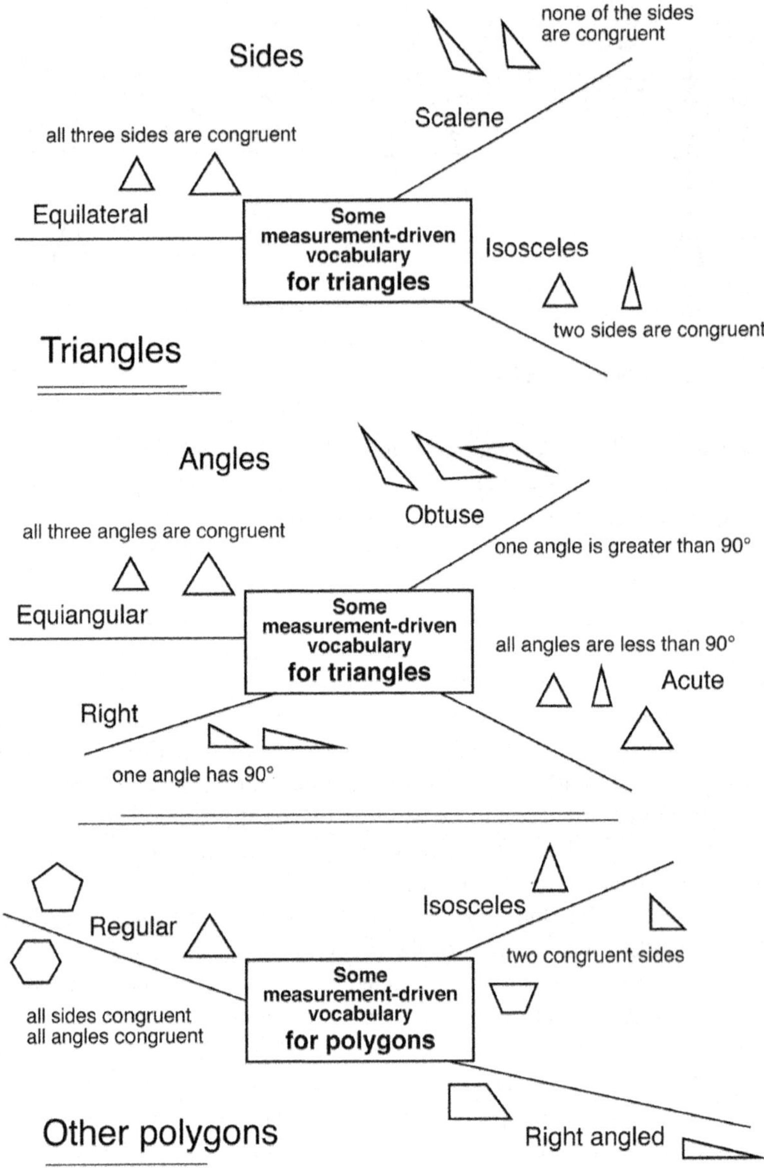

Figure 3.5   **Measurement-driven webs for triangles and other polygons.**

three sides and three angles, the measures of the sides and angles are limited. The total angle measures can never be over (or under) 180° and two side lengths can never add to equal the third side length. Our geometry guru Euclid established these ground rules for the side measures triangles and stated, "in any triangle, the sum of any two sides is greater than the remaining one."[3]

As long as the triangle sides combine in this way, the lengths of these sides can be as long or as short as your child's heart desires, or the next geometry problem requires. Two sides of a triangle can be congruent, even three sides can be congruent, or none

Prove It to Yourself:

Cut some straws or pipe cleaners into three different lengths: 5 inches, 3 inches, and 2 inches. Measure twice and cut once. Try to make a triangle with these lengths.

Can you make a triangle with these lengths?

Cut three more different lengths, measuring them so that the sum of the two straws or pipe cleaner measures is just as long as the third measure.

Can you make a triangle with these lengths?

Answer: No.

Figure 3.6　Task Box: Prove It to Yourself.

of the sides can be congruent, provided that the sum of two of the triangle's side measurements will never equal the length of the third side. The total angle-measure limitation for triangles is more fully explained in Chapter 5 along with other two-dimensional shapes in Flatland.

All of those names of different triangles and quadrilaterals come from the combinations of angle measures and the side measures. In fact, all polygons can have two measure-driven descriptors, one for angle and one for sides, except when the term "regular" applies because the term "regular" takes care of both angles and sides at the same time. Some of these combinations are shown in Figure 3.7, while other polygons include these angle and side combination characteristics as their properties.

The more frequently used polygons, the triangle and the quadrilateral, have their own independent name when they are regular. For triangles, it is known as an equilateral triangle and for quadrilaterals it is known as a square. An equilateral triangle *is* a regular triangle and a square *is* a regular quadrilateral, but the terms used on most tests are "equilateral" (or equiangular) triangles and "square." If the polygon doesn't include "regular" as part of its name, then it could be irregular.

Instead of the traditional explanations for shape definitions, let's try working with polygon definitions from a different perspective. You and your child can examine these shapes from a dynamic perspective—how they got to be what they are from their angle measures and their side orientations and measures. For example, if a 90° angle is in a triangle or quadrilateral shape, then that angle pretty much controls the rest of the arrangements of the sides.

By the same token, if any pair of sides in a quadrilateral has a parallel arrangement, then the other sides are likely to be controlled in how they are arranged as being congruent, parallel, or both. With the other polygons having more than four sides, the control of perpendicular or parallel sides and one 90° angle has less effect on the rest of the sides and angles. Generally speaking, the angles in a polygon determine how the shape looks, and the sides in a polygon determine how large the shape is.

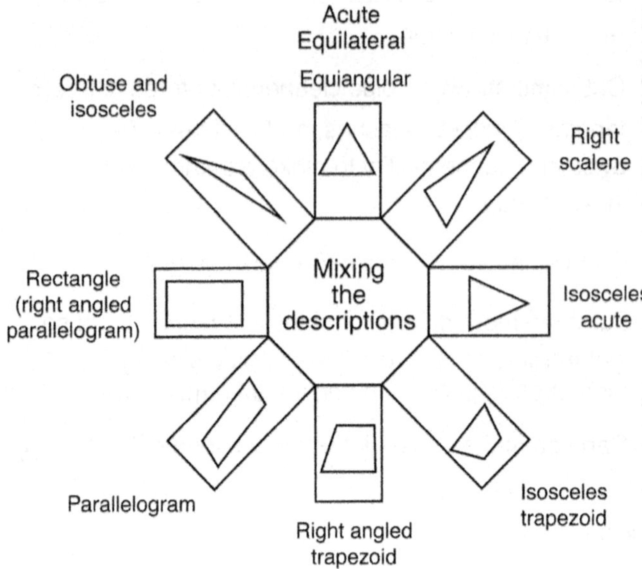

Figure 3.7   Octagon web for mixing arrangement and measurement descriptions.

KEEP IN MIND

Learning new vocabulary is not easy, even for Kermit. The language in geometry includes descriptive words for angles, sides, and number of sides in a shape. Realizing that the clues for the shape names is embedded in the name can help. Anytime you can categorize information—or as others call it, "chunk it down into smaller pieces"— using your own interpretations and perspectives will make learning new information make more sense, and fit into what you already know. Draw your own web diagrams to help make your connections make more sense.

Ask Yourself:

Could a triangle have parallel sides?

Could a triangle have more than one 90° angle?

Draw or sketch some triangles to test your ideas.

Figure 3.8   Task Box: Ask Yourself.

NOTES

1. Armstrong, *Multiple Intelligences in the Classroom*, 147.
2. http://aleph0.clarku.edu/~djoyce/java/elements/elements.html
3. http://aleph0.clarku.edu/~djoyce/java/elements/elements.html

Part II

ORGANIZATION

The following is a conversation between two inhabitants—one from *Flatland* and the other from *Spaceland*. The two are involved in an inquiry about the existence of each other's world.
I (who lives in Flatland and speaking to Stranger after having just met him): Would your Lordship indicate or explain to me in what direction is the Third Dimension unknown to me?
Stranger (who lives in Spaceland is answering I's perplexed question since I has never seen anything but flat inhabitants): I came from it. It is up above and down below.[1]

—Edwin Abbott

Whether you feel as perplexed as "*I*" feels or have a larger view of the space world as "*Stranger*" does, the idea of being able to organize all these different kinds of shapes into categories can bring a sense of calm—no surprise pop-ups of unfathomable shapes, just organization. In this three-dimensional world where *Stranger* and the rest of us live, all of the geometric shapes, even the perplexing ones, are built from the shapes in the previous dimension, and yes, from that zero dimension as well. Interestingly, we first encountered these dimensions in that place value organization for numbers.

Although *Flatland* was written as a political satire, the book's inhabitants in Lineland, Flatland, and Spaceland are described with clearly identifiable geometric properties. Each inhabitant living in a different dimensional land is unable to "see" each other across space boundaries. The inhabitants of Lineland in Chapter 4 live in a one-dimensional space measured only by lengths, the inhabitants of Flatland in Chapter 5 make their home in a two-dimensional plane generated from lines, and the inhabitants of Spaceland in Chapter 6 exist in a three-dimensional space established from two planes, usually perpendicular.

While the topics in this chapter cannot help the inhabitants of Lineland, Flatland, or Spaceland get to know each other across boundaries, they will help you and your child make more sense of all those different dimensional measurements of shapes in the Euclidian geometry that your child is learning in school. Yes, there are other geometries; but Euclid's geometry is the one that your child is probably learning in math class.

All of the geometry formulas and other information that you may have felt compelled to memorize are organized in this chapter within dimension categories so that your child will be able to recognize and even use the simpler shapes to figure out the more complicated shapes. For example, perimeter and circumference from Lineland are expressed by length, yet are made up of points from the zero-dimension space and have no meaning until they wrap around a two-dimensional shape in Flatland.

In the Flatland territory dimension, the area that can cover the interior space of two-dimensional inhabitants is calculated by using one-dimensional lengths and widths. Spaceland inhabitants in the three-dimensional shapes have volumes that are generated by using those two-dimensional areas in Flatland. These two-dimensional shapes wrap around three-dimensional shapes in those "going above or below" planes in Spaceland. Talk about crossing boundaries!

The elegant details of properties and formal shape definitions are left to Euclid while you and your child can focus on the organization of the shapes. Shapes and their component parts are easier to talk about when described with measurements, so several concepts involving familiar measurement of shapes are included in this chapter. These are not exhaustive explanations, just sufficient for general understanding. Don't forget graphs. Graphs are a visual and geometric version for showing trends of numbers in data.

Geometry isn't finished and it isn't limited to a fixed set of formulas, but it *is* organized. There is still plenty of space left for exploration. The only thing that limits your child is imagination. If your child stays within Euclidian space, then they must abide by the boundaries determined by Euclid. Of course, your child, like Riemann or Lobachevski, is invited to venture out into uncharted thinking but, also like Euclid (flat plane) and Lobachevski (hyperbolic plane) or Riemann (curved plane), will need to develop their own new set of definitions and properties as well to define his own boundaries.

Even the mathematicians would like to nibble at the forbidden fruit, to glimpse what it would be like to . . . to slip a moment into a fourth dimension.[2]

—Kasner and Newman

NOTES

1. Abbott, *Flatland*, 70.
2. Kasner and Newman, *Mathematics and the Imagination*, 124.

Chapter 4

Ribbons, Rolls, and Rulers for the First Dimension

It need scarcely be added that the whole of their horizon was limited to a Point; nor could any one ever see anything but a Point.

—Description of a Linelander, an inhabitant of Lineland[1]

The inhabitants of the imaginary Lineland, like the rest of us, are sometimes limited by what we see. Linelanders can see only a point of their fellow one-dimensional countrymen. They recognize each other through an array of properties that they *can see*. Leave the elegant details of properties and formal shape definitions of these one-dimensional geometric inhabitants to Euclid while you use some of his definitions to help organize the shapes and terms in today's geometry class. Shapes are easier to talk about when described within their "environment," so the linear-anatomy inhabitants in this chapter are described by where and how they "fit into" their surrounding shapes.

Lines may look easy enough, however, they can prove to be confusing and difficult to understand when they "change their shape" like a shapeshifter, and look like an inhabitant of Flatland. For example, how can a round circle be straight? How can a "fenced-in" closed polygon shape be straight? These are only two of many questions that children face with those one-dimensional geometry definitions and relationships. The answers to these questions, and other questions, follow in this chapter. The Lineland inhabitants cannot see the inhabitants of Flatland any more than your child can cope with these strange changes.

TRANSITIONING FROM ZERO DIMENSION TO ONE DIMENSION

Is anything really zero dimensional? Yes, a point has zero dimension, no length and no width. Euclid's *Elements* started with the first definition that specifies "a point is that which has no part." Later he further expanded on the point idea with the definition that "the ends of a line are points." Since the line has length in one dimension, it would be reasonable that the ends of that line have no measure, no dimension to measure, no parts. Are Euclid's details necessary to help your child with spatial sense? Maybe not initially, but knowing how these dimensions coexist can help your child understand the measuring of those line segments, rays, and heights.

The one-dimensional shapes are lines, rays, and line segments, or as Abbott called them, the inhabitants of Lineland. They come in many sizes with many names: circumference of a circle, sides and perimeters of polygons, lines, line segments, rays, heights or altitudes of shapes, lengths, widths, diameters, diagonals, radii, radians, chords, secants, and tangents. Yes, there are more, but recognizing all of these as line measures (except the unending lines) will give you a heads-up for identifying all of the other linear images that are just waiting for your child to find them.

Euclid acknowledged that lines had to come from somewhere, so he defined the line as "a breadthless length" and "the ends of a line are points."[2] When our line inhabitant in Lineland has endpoints, then it is referred to as a line segment—something that is measurable because it has a beginning and an end, quite reasonably called "endpoints." An unrestrained line, on the other hand, does *not* have endpoints so it really isn't measurable, yet it is still one-dimensional. A ray has only one endpoint, allowing the other end to go on and on, so it too really isn't measurable. A ray has a beginning but no end in sight. Welcome to Lineland!

THE ROLLS OF CIRCUMFERENCES AND PERIMETERS

The geometry terms and measures in one-dimensional space have names that all refer to lines or line segments. A few of these line segments as linear measures can be confusing because they do not look like one-dimension unless you roll them or trace them as track lengths. Two particularly difficult examples to see as one-dimensional lines are circumference and perimeter because *they don't look like lines* when they are in a shape. It could be that your child does not really "see" circumference as a linear measure (how could they, it is round, isn't it?) or perimeter as a linear measurement.

Because circumference and perimeter don't look like one-dimensional lines, their formulas can be difficult for your child to remember. They may take it on faith, but do they really buy into it? You can recognize this by their response when you ask for the perimeter and they quote the area formula, or you ask for area and they reference the four sides. Neither perimeter nor circumference measures exist until each has encased its respective shape in space. Simply put, a one-dimensional measure must stay in one-dimensional Lineland units.

As for the polygons and their perimeters, your child can use fold-out rulers that carpenters used way-back-when, pipe cleaners, straws, Anglegs, or Exploragons (Figure 4.1) around a polygon's rim or edge. All of these tools will allow your child to get hands-on experience and a feel for making a polygon. When the shape is unfolded, the linear perimeter reveals the straight line-ness and becomes a line segment *that actually looks like a line* that they can measure, not just take on faith that it is "the outside" of the shape. Using separate straws and sticks to make a polygon supports the idea that a polygon *is made with straight line segments.*

Another issue arises when the straight perimeter curves into a circle circumference. You will need yarn pieces or anything else that bends (Figure 4.1), like pipe cleaners or ribbons, to form the roundness of a circle. That stretched-out yarn, or pipe cleaner

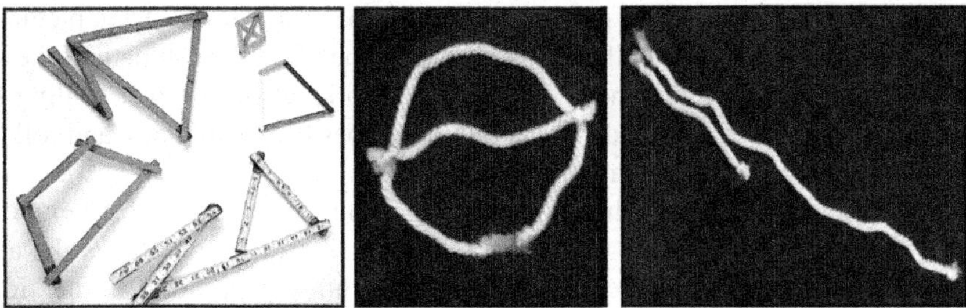

Figure 4.1 Shape maker foldout rulers and yarn circle.

or ribbon, makes the circumference look smooth and, yes, very much like the line-segment-made perimeters. The proof is in the making so that "seeing is believing."

Neither perimeter nor circumference measures exist until each has encased its respective shape in space. That is why the circumference and perimeter encased shapes need to be unwrapped so that they look like the one-dimensional line that they are supposed to be. When this happens, the confusion between line and square measurement units dissipates and the lines don't look like area unit squares anymore. The dependence on "formulas" tends to fall away, and the kinds of units in the measurements start to make sense. One-dimensional units are used in Lineland and two-dimensional units are used for Flatlander areas (see Chapter 5).

If you do not have access to these folding tools, you could show your child how to "roll" a shape, marking the image of the roll very carefully along the edge of a ruler on a path like the ones in Figure 4.2. Being able to *mentally* remove a perimeter "off"

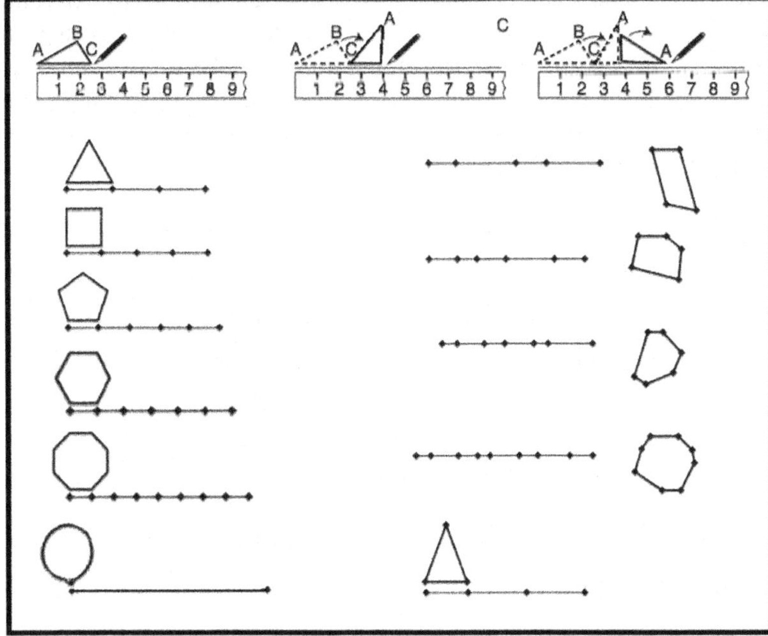

Figure 4.2 Polygon rolls for one-dimensional perimeters.

of a page in a book and make it into a long line can be problematic. A flat picture on a page in a book doesn't roll. Rolling a block can do the trick; the roll path of the shape gives the perimeter a better linear feel. Tracing works also, however, rolling has a better "track record." Your child can cut out the shapes from cardboard and roll them, or use some shapes from their assortment of games.

Roll a variety of shapes, being careful to mark the beginning and the end on the track. For polygons, mark the end *of each side* so that the track shows the appropriate number of sides. As the shape track rolls show in Figure 4.2, the sides of regular polygons will look quite different from irregular polygons. This activity can easily convert into a "guess my shape by the roll" game or "can you draw this shape from its roll?" game after understanding how the sides of the shape have "tracked" onto the roll. This rolling idea also helped a fifth-grader named Ned to recognize lengths and widths as edges needed for area (instead of just saying outside the shape).

After rolling all of those polygons, make sure to include a circle's roll for comparison. There's a very good reason why car tires are round instead of the bumpy ride from a polygon-shaped tire. The trick of that smooth ride is in the roundness. Even when it is highlighted with a color, a picture image of a circumference belies its "one-dimensional-ness." That flat picture doesn't look or feel very linear to your visual-spatial learning child. Rolling a circle and carefully marking the start and stop places, like rolling a perimeter, has a more natural feel to it, and even looks like a bicycle track line in the snow.

Any rounded shape brings into play one of those transcendental and irrational numbers with a decimal format that never repeats and never ends. Pi (π) is that irrational transcendental number for circle circumferences and other rounded shapes. Mathematicians and computers are still calculating pi to determine if the decimal pattern will ever repeat, and if it does, then π will be recategorized and reorganized and placed in the section for rational numbers. After all of the work that mathematicians have done with pi over the centuries, this change in category will probably not happen. More about these pi (π) comparisons with circles are explained in Chapter 8.

Try This:

Take turns. Choose a secret shape block. One person rolls a polygon behind a screen so that no one can see the shape or the roll. After the screen is lifted, the other person has to guess the polygon just by looking at the roll pattern.

Make this into a game of "guess the shape from the roll pattern."

Figure 4.3 Task Box: Try This.

If your child must use a formula for perimeter or circumference, then certainly encourage him to use the addition of all of the sides for perimeter rather than memorizing-by-rote a multiplication that does not make sense to him. If memorizing was an unpleasant experience for you, then you can be sure that it will affect your child the same way. That is, unless your child already thinks like a mathematician—at which point you need to allow your child to explain things to you.

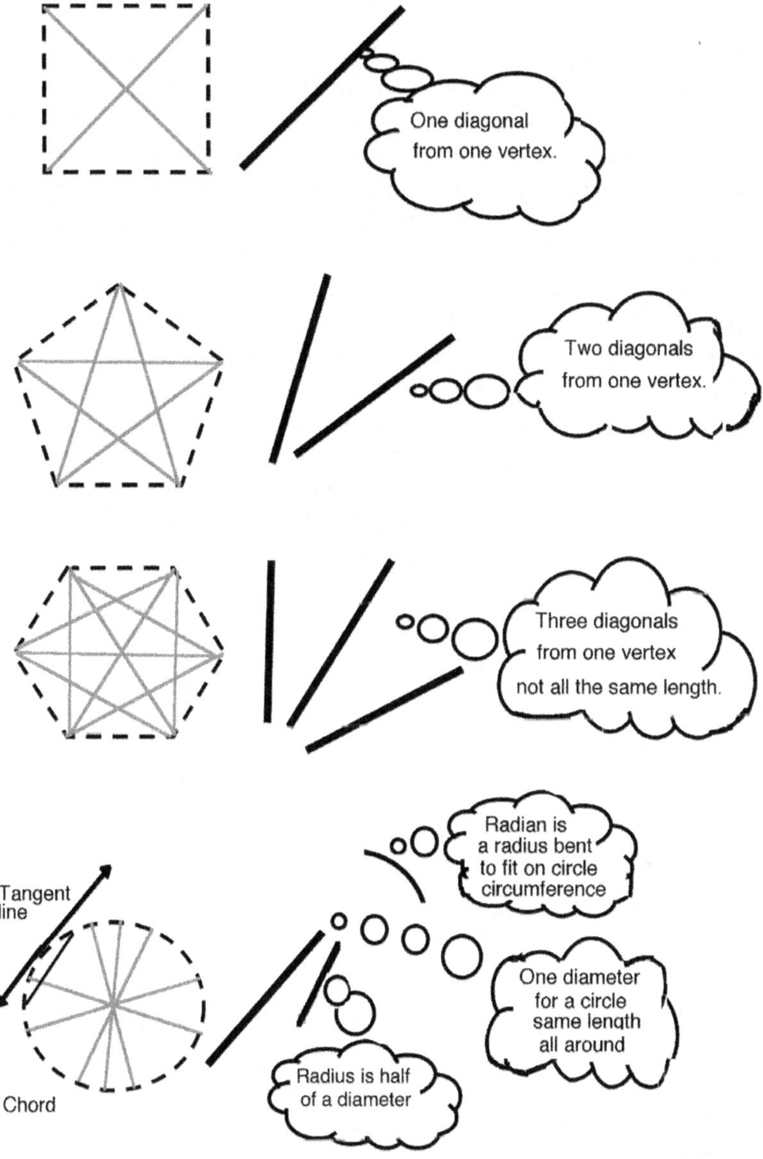

Figure 4.4   Diagonal and diameter pull outs.

DIAGONALS, DIAMETERS, RADII (RADIANS),
AND MORE STRAIGHT PIECES

Diameters need a circle in order to be a diameter and diagonals need a polygon to be a diagonal. Also, a radius, a radian, a chord, and a tangent need a circle for their identity and measurement. Despite the fact that they need a two-dimensional shape in order to get their name, these are still measured as a one-dimensional length. Chords just go through a circle from one location on the circumference to another, very much like a shortcut across; the diameter of a circle is a specific chord that goes through the center of the circle so it takes the long way across. The tangent line to a circle is a line that hangs on tightly to the circle at one point.

To extract the diagonals, diameters, radius (radius can be a radian when the radius is copied and curved so that it marks an arc on the circle circumference), and any other linear element in a shape, your child can use tracing paper to trace the specific line segment length and move it away from the shape. The polygons in Figure 4.4 show diagonals that have been removed from one vertex so that their one-dimensional measure is more obvious. The circle in Figure 4.4 has a few more one-dimensional relationships, especially diameters, than the polygons. Circles, by their very nature of being round, have a special nobility status in Flatland and in geometry.

Will ovals roll too? The oval's formal math name is "ellipse." Because it isn't rounded like a circle, the ellipse has a shape like the one shown in Figure 4.5 and has two diameter-like segments instead of one. These two quasi-diameters are not the same length, so they need new names: major axis and minor axis. These new names make sense, since one is clearly longer than the other. Yes, there is a way to actually calculate the lengths of these axes, but for now let's just roll them to recognize their one-dimensional circumference and acknowledge that they, too, would need pi to calculate their area as well as their circumference.

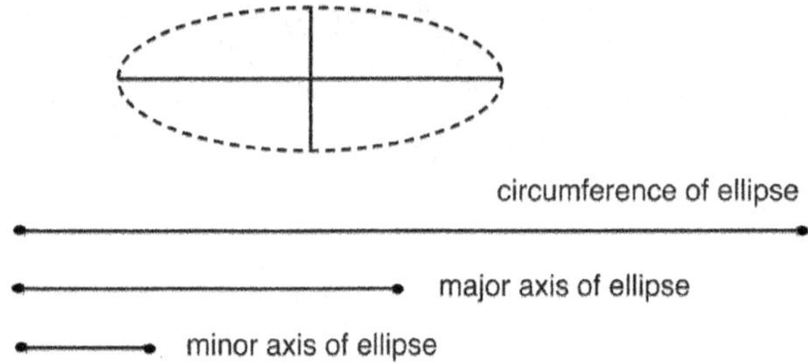

Figure 4.5   Ellipse with one-dimensional circumference, major and minor axes.

HEIGHTS, ALTITUDES, APOTHEMS, AND LEGS

Does your child know that the altitude or height of a triangle, or any shape, does not have to be inside the triangle? In fact, a triangle can have three different altitudes! All that is required for an altitude or height is that it be perpendicular to the base of the shape (or an extension of that base), just like standing next to the door to measure your

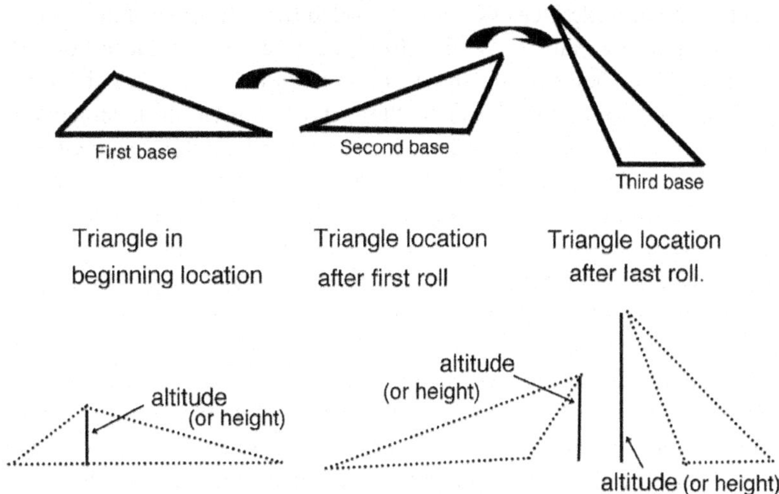

Figure 4.6   Roll the triangle to find different bases and matching altitudes (or heights).

own height. Altitudes are primarily useful only for triangles and quadrilaterals. They are also used when working with polygons with more than four sides, but are usually referenced within a triangle as part of the shape composition. Keep reading, because this will make more sense as you continue.

Imagine rolling a triangle like the track rolls in Figure 4.2 for a rather bumpy ride, like the sample triangle in Figure 4.6. Each side will take turns being a base of the triangle. To show one triangle roll, imagine the triangle on the left rolled to the right until it lands on the adjacent side (in math translation terms, this is called a rotation, explained in Chapter 7). Roll it again until the triangle lands on the third side. Each time the triangle rolls to a new base, a new height (altitude) is created to go with that particular base.

Prove It to Yourself:

Draw a triangle, any triangle will do.

Measure the base. Draw a height for the triangle that goes with that base. Measure the height.

Roll the triangle to another base. Measure that base. Draw a new height for that new base and measure it.

Multiply base times height for both triangles.

Compare the two areas. Are the areas the same?

Figure 4.7   Task Box: Prove It to Yourself.

To identify the altitude measure for each rolled triangle, your child would measure how high the tip of the triangle vertex is from the base. Three rotations of this triangle make three *different* heights and three *different* base lengths, yet the formula that multiplies the base times the height of these three triangle arrangements maintains the same area *because it is still the same triangle.* If a height perpendicular to the triangle's base isn't there, then you need to draw one (the Math Aficionado expression is "drop a perpendicular") from the highest vertex to the base (or extension of the base if the dropped perpendicular height falls outside the triangle).

For right triangles and rectangles, it is easier to recognize the height or altitude because the height is already there as one of the sides (Figure 4.8). For a right triangle, one of the legs is the height, and since all of the angles in the rectangle have 90°, any of the sides could serve as heights. For non-right-angled shapes, you need to draw the height measure if it isn't already there, just as you did with triangles. As a general rule, only change the orientation of a shape if it serves your problem-solving purpose or helps you answer the question. Remember, a specific right angle and height measures do not change when the shape itself is in a different orientation.

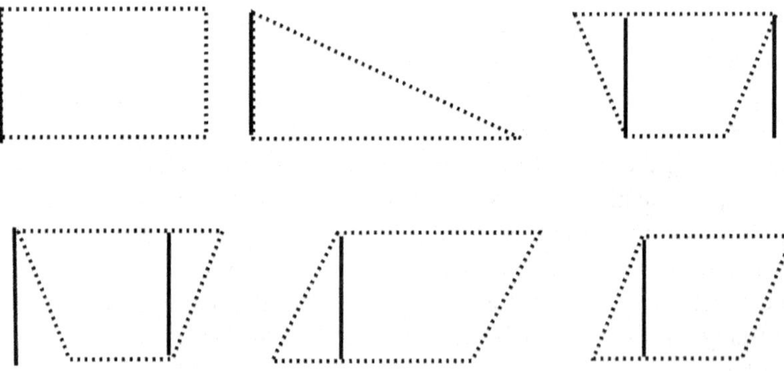

Figure 4.8   Polygons with altitudes (heights) marked.

By the time your child is using polygons with more than four sides, they will be using altitudes of shapes that are part of the polygon's composition. In the case of the regular pentagon and regular octagon in Figure 4.9, the height of that internal triangle gets a new name, "apothem," because it is contained within the regular polygon and does not extend from the top of the entire polygon. That apothem serves as the height *for the triangles* so that the areas are calculated and added together to represent the area of the regular polygon. We can thank the Greeks for the word "ancestry": "apo" and "thema."

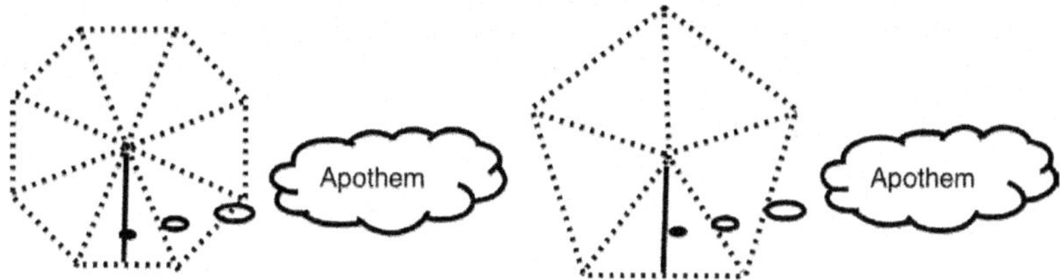

Figure 4.9　Apothems in octagon and pentagon.

KEEP IN MIND

Generally, when a math problem requires calculating with altitudes or heights, the context of the problem will provide the height measure or enough information to be able to figure out the actual height length. More about these connections are described in Chapters 5 and 10. Like many children, middle-school student Lizzie from Chapter 1 needed to see a specific orientation for a right triangle before she would agree that the angle was a 90° angle. If it helps to see the shape on a page in a different way, rotate the book and ignore any doubters who question your motives.

NOTES

1. Abbott, *Flatland*, 56.
2. http://aleph0.clarku.edu/~djoyce/java/elements/elements.html

Chapter 5

Angle Swings and Area Squares
in Flatland

Imagine a vast sheet of paper on which straight Lines, Triangles, Squares, Pentagons, and other figures, instead of remaining fixed in their places, move freely about, on or in the surface, but without the power of rising above or sinking below it.

—Description of a Flatlander, an inhabitant of Flatland[1]

The Flatlander triangles, squares, pentagons, and other two-dimensional shapes move about freely in this chapter. The inhabitants of Flatland are characterized by the number of their sides and then by the congruency of their sides. As a reminder from Chapter 1, when all sides of a polygon are congruent (the lengths of the sides are equal), the shapes are called "regular." When the polygon sides are not congruent, then the polygons are called just by their polygon name, or they may be described as "irregular," or possibly "similar with proportional sides." In Flatland, as in math lessons, the regular polygons have a privileged status and tend to run the show!

Too often, formulas take control of the math so that children don't really get the feel for the relationships defined by the formulas. These formulas resulted from patterns and predictions about how the shape pieces work together. As the rhombus decompositions examples from Chapter 1 are described as a formula in Chapter 12, the formula for the area of a rhombus, any rhombus, evolved from decomposing the area of the rhombus. After the rhombus was disassembled, it was reassembled, using a little rearrangement, a lot of flexibility (assisted by the commutative property for multiplication), and knowledge about averages to finish the job.

Taking apart, reassembling, and being able to recognize how all of the geometry pieces fit back together in both geometry and generalized arithmetic (known as algebra) are all part of the essential critical thinking skills involved in having spatial sense. Taking apart these two-dimensional, very manageable shapes "on or in the surface" of Flatland prepares your child for other spatial skills in other dimensions.

ANGLES ARE AT THE ENTRY GATE TO FLATLAND

Angles control two-dimensional shapes. As soon as your child leaves the one-dimensional length and begins an angle "swing" in another direction, then they have entered the second dimension and have moved into Flatland. Angles have a great deal of control over the shapes that they are in, and they certainly affect those shapes' calculated areas. Thales of Miletus, who hailed from fifth century BC, knew that if he was given two angles and the side in between, then he could generate a triangle *just from that information.* More information about Thales and angles and triangles is described in Chapter 10.

Drawing an angle in Flatland requires two one-dimensional lines (or segments or rays) in two different directions. An angle starts on a one-dimensional ray (or a line or a line segment) and swings (rotates is another way to envision it) around to another ray (or line or line segment) for the two-dimensional layout. Angles can swing all the way around to the starting ray in a circular fashion, ending up at 360° for the first go-round of the circle. The now familiar 360° angle measure for the circle is credited to Hipparchus, a mathematician-astronomer in the second century BC who likely got the idea from Hypsicles, who had divided a day into 360 parts.[2]

The angles in a shape can and do control the orientation of other sides of a shape, except for the two sides that form the angle. For a triangle, any given angle controls the opposite side of that triangle. For quadrilaterals, an angle will influence the two sides opposite that angle. For polygons with more than four sides, the impact is less dramatic but still evident. More explanations about angles and their influence on side lengths and total angle measures in polygons reappear in Chapter 8 and Chapter 11.

Since an angle is measured as a swing or rotation, otherwise called an arc, a straight ruler cannot measure it because it is rigidly straight (from Lineland territory) and cannot measure the space in a rotation between two lines in Flatland territory. Your child will need a new tool that can measure a swing around angle or arc like the ones shown in Figure 5.1. That tool is called a "protractor," a term first used in the seventeenth century and probably derived from medieval Latin. A protractor can measure the full 360° swing of a circle, although many protractors show a scale for only a half circle and rely on the user to duplicate the half-circle for the full revolution.

A good way to get the "feel" of an angle is to use two different color circles or paper plates, cut the circles along a radius, and then slide the two circles together so that they rotate to make different angles. Kenzy made the circle "angle" model in Figure 5.1 with help from her mom. Kenzy was a visual and kinesthetic learner in the fourth grade and was having a great deal of trouble, like many children, with the idea

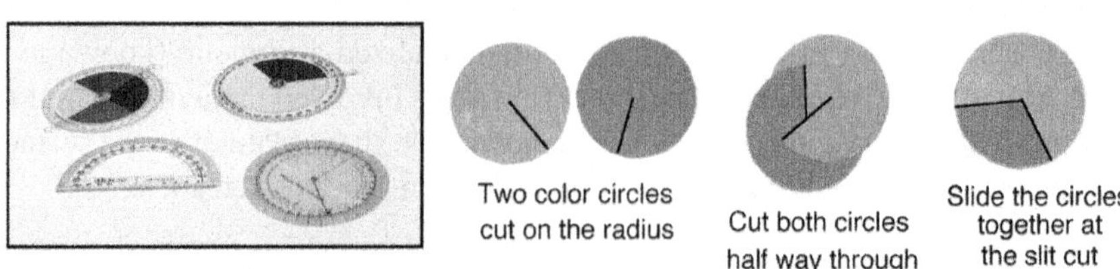

Figure 5.1   Assorted protractors and Kenzy's angle maker.

that a ruler couldn't measure an angle. Just the experience of cutting the circles and making the two circles rotate gave her a better idea about how rotating circles make angles and sections. This model also doubled as a visual for fractions.

A triangle's side and angle limitations were explained in Chapter 3. In fact, the total of all three of the angle measures in a triangle is indeed limited to 180°, a fact your child may have memorized for a test. Like high-school student Theo from Chapter 1, your child may not have realized that the angles, when rearranged, would compose a straight line. In fact, your child can cut out any triangle, tear off all three angles of that triangle, align them next to each other around a central dot, like the three triangles in Figure 5.2, and then they will see the same 180° straight-angle straight-line result. It may have helped Theo if he had seen this in his earlier years.

Janis was bothered by the idea that different triangles could have three different angle measures. She said, "But I thought *all* triangles *only had* 180° in the triangle." She was confusing the inflexibility of the *total degrees* in a triangle with the flexibility of having *three different angle* measures that *could be added to get the total of 180°* in a triangle. Developing spatial sense had not been part of Janis' math development. Her experience had been one of memorizing, not flexibility. Your child, and Janis, can get to 180° in many different ways, since different angle combinations make different triangles. Flexibility with compositions to the rescue!

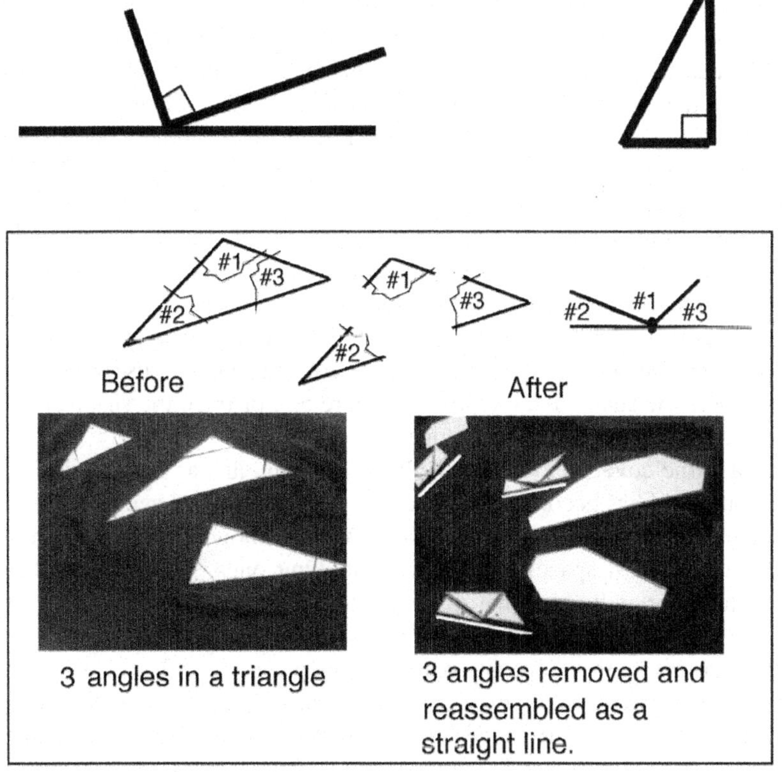

Figure 5.2 Sum of angles in a triangle (and straight angles) will always total 180°.

A quadrilateral's side limitations were referenced in Chapters 2 and 3 and will be explored in later chapters; however, the total measures of the angles of any convex (memories of Chapter 1) quadrilateral will never be greater than 360°. The term "convex" must still be included to ensure that a quadrilateral's angles will stay under the 180° straight-angle mark. The total of all three of the angle measures in a triangle is indeed limited to 180°, so if any quadrilateral is composed of two triangles, you will never have more than the 360° total. Connecting this idea of adding angle degrees in triangles to figure out angle degrees in other polygons comes up again in Chapter 11.

AREAS AS SPATIAL ARRANGEMENTS

Now that the angle "gatekeeper" has allowed two-dimensional shapes into Lineland, more two-dimensional shapes are formed and their areas are measurable. The rectangle area formula, base times height, seems to be the easiest formula to remember, so wouldn't it be delightful to be able to use one rectangle formula for lots of area calculations? One formula that can be used with many applications—who wouldn't like that? This area formula for rectangles also applies to parallelograms, provided the appropriate height is matched to the related base. All that is needed now is to identify the rectangle or parallelogram!

A shape's area represents the coverage within the boundaries of the shape perimeter. This coverage concept began with base ten blocks back in the first grade, when children are making a ten or a hundred block with their ones cubes for place value, but at that time it is not referred to as area. You or your child may refer to area as the inside measure, a reference that is reasonable unless the partner reference is that "perimeter is the outside," another shortcut that is confusing when the "outside" refers to the entirety of space external to the shape. Katrina resolved her confusion by drawing her own pictures with her own notation, shown later in this chapter.

Counting the squares for an area inside a rectangular shape is reasonable as long as the area has complete squares. For rectangles, all of the squares are there, not interrupted and not ever counted twice. However, the triangle, parallelogram, and the circle can present a challenge since all of the squares in the area are not immediately apparent. These particular challenges are described later in this chapter. Showing areas as coverage can be shown with same size tiles to cover a shape, or use a transparent grid paper to cover the shape like the ones in Figure 5.3, or trace the shape and color the square pieces inside the traced line edge.

An area is the coverage; the area isn't the formula, just as the map isn't the territory. If your child has become entrenched with formula-based geometry, they will be surprised to learn that they can cover spaces without knowing a formula. In fact, your child can cover most spaces, even strange-looking ones, with smaller shapes that, when put together, will cover that space. Children are delighted to learn that they can also figure out the area of a larger encompassing shape and then subtract some cutaway areas or cut up the shape and rearrange the pieces.

The rectangle area formula even applies to triangles after the area is divided into two halves. As Ian Stewart suggested in Chapter 1, all polygons are composed of triangles, so since quadrilaterals are polygons with four sides, then they, too, can be composed of and decomposed into triangles. Altitudes and heights were described in

Figure 5.3   Areas of shapes measured by covering them with squares.

Chapter 4, so now you can use those ideas to calculate areas. But what about all those other shapes? Will base times height work with them too? Yes, but you and your child must be flexible with rearranging, composing, and decomposing the shapes in order to successfully apply this basic formula to just about any shape.

Starting with areas of rectangles as an easy example, those base ten blocks used for place value in the first and second grades can now be connected to using place value to spatially organize the numbers in an area calculation. The arrangement itself will apply to all kinds of numbers (integers, fractions, decimals, and even the algebra unknown "x"), and is easier to recognize when blocks are used to arrange the multiplication of "18x19" in Figure 5.4 with the base ten blocks (or any cubes or squares).

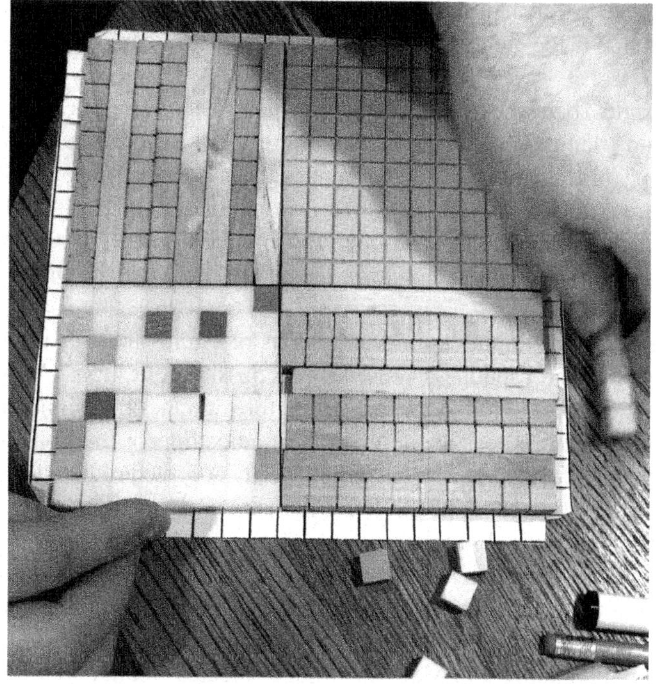

Figure 5.4   Two digit multiplication with base ten blocks represents area.

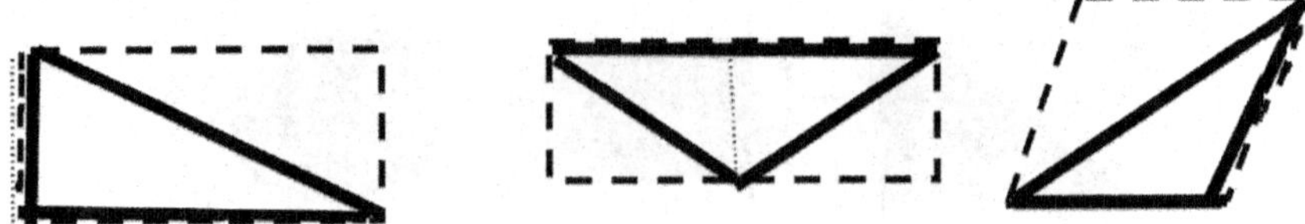

Figure 5.5   A triangle area is one-half of its matching rectangle area.

Now let's talk about how the triangle fits into a rectangle (and a parallelogram, too). A triangle is part of a rectangle (or parallelogram) as long as the triangle and the rectangle (or parallelogram) share at least one side, preferably two sides. The decision to draw a parallelogram or rectangle is *your choice*, whichever makes solving the problem easier. The three examples in Figure 5.5 show triangles as part of parallelograms and rectangles that, with some imagination and rearrangement, show how the triangle area is one-half of a rectangle or parallelogram area. The heights are marked with a dashed line.

Because the sides of the rectangle are appropriate for base and height, then the "base times height" formula for calculating a rectangle's area works every time. Finding the area of a parallelogram is a different story because the side of the parallelogram is not always representing the height. Since every parallelogram *does have a height*, your child may be required to draw one just as they did for the triangle height in Chapter 4.

In each case, the triangle area is always half of the surrounding rectangle or parallelogram area. If you want to impress someone about the area of a triangle, then you could use our fifth-century mathematician-astronomer Aryabhata's definition, "tribhujasya phalashariram samadalakoti bhujardhasamvargah"[3] (if indeed you can pronounce it!). The translation is said to be "for a triangle, the result of a perpendicular with the half-side is the area." Good thing the twenty-first-century expression is easier to express.

Non-rectangle shapes like trapezoids are also "decomposable" into triangles or rectangles and triangles (see Figure 5.9). There's also a formula for calculating the area of a trapezoid, if you are not that interested in decomposing shapes. The result is the same area, regardless of which strategy you prefer. That formula for the area of a trapezoid requires a calculated number, or "figured algebra" as Germain describes it in Chapter 12, that you will recognize as a calculated average of the two opposite parallel side measures.

Circles present a different conundrum. Areas of circles must deal with the roundness of the edge. The roundness presents two possible questions: Why do we have to adjust for roundness? How does one adjust for roundness? Archimedes is credited with developing the first reasonable method of adjusting for this roundness by using regular polygons. For now we can use the generally accepted rounded decimal number 3.14 for our calculations with pi and leave the exploration about "why" for Chapter 8 and special comparisons of numbers.

That circle connection to a rectangle in order to be able to use the rectangle or parallelogram area formula will require cutting the circle into congruent wedges (or pie pieces) and then rearranging these pie pieces into a shape that begins to look like a slightly bumpy parallelogram. More cut wedges creates smaller wedges and the reassembly will begin to look like a rectangle. The concept of a limit from calculus study

Figure 5.6   Task Box: Prove It to Yourself.

allows this kind of approximated transition from a circle to a rectangle shape. All you need now is the general idea to help explain this circle-to-rectangle idea to your child. A photo and the algebra description are in Chapter 11.

Circles also can be covered with squares to show area, just as polygons can have square area, but covering a circle is a bit trickier because of the rounded circumference. Covering the circle with four smaller squares as shown in Figure 5.7 will leave some extra noncircle scraps around the edges. The total area of the four smaller squares is too much to be the area of the circle, so the area of the circle has to be less

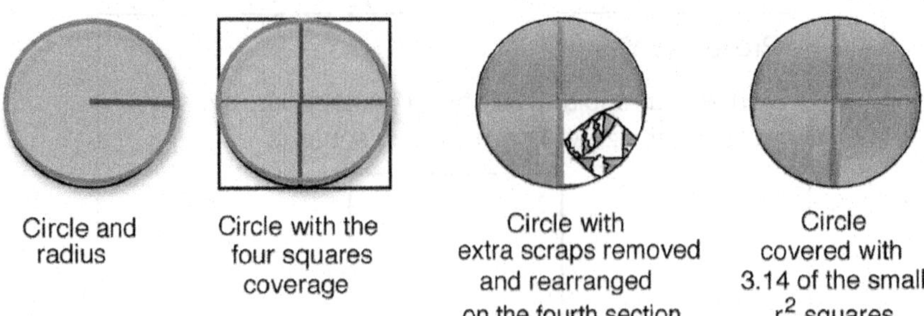

Figure 5.7   Circle area from covering with squares then trimming the extra scraps to complete the fourth circle section.

than four of these squares. It is not an accident that it takes a little over three small squares to cover the circle area, the same 3.14 number that is generally used for pi. More about pi and its relationship to circles is explained in Chapter 8.

To see (with your hands) how the circle gets covered *without the extra scraps*, cut off the extra scraps and tear them into smaller pieces, then use these scraps to cover that fourth piece *of only the circle*. They *almost* cover that last circle section. In fact, the scraps cover almost 86 percent of that last fourth part of the circle. This rearrangement confirms that three square areas (base times height or "r times r" or $3r^2$) covers too little of the circle and four squares (or $4r^2$) covers too much. You will need an additional 14 percent more of *a small radius square* to complete the circle area (or 3.14 or pi). The scraps are rearranged, so now you are left with only the circle area.

Get the idea? Once your child believes in general that the base multiplied by the height (or length times width) is area for a rectangle, then they are empowered to mark off any strange-sided shape with parallelograms or rectangles or triangles and then calculate the area of that shape using the rectangles or triangles that are used to compose that shape. Or they can cut up the shapes, rearrange them, and then calculate the area. Of course, there are formulas that apply to calculating many of the strange shapes, but all of them are generated from the basic idea of length times width.

PERMISSION TO SURROUND AND REARRANGE SHAPES TO FIND AREAS

Continuing with the idea of using familiar shapes to help find areas of other less-familiar shapes, use a rectangle or a parallelogram to surround (encase or draw around) the shapes. An unfamiliar shape can be tamed and surrounded with a parallelogram or rectangle. Again, the decision to draw a parallelogram or rectangle is *your choice*, whichever makes solving the problem easier for you. For example, the shapes in Figure 5.8 were encased by either rectangles or parallelograms that leaves leftover pieces to trim. Those extra areas are calculated and subtracted (using Chapter 4 information) so that the area of the original shape is the end result.

The choice is yours to (1) choose whether to use a rectangle or parallelogram strategy and (2) choose whether to surround, rearrange, or cut away as your strategy of choice. Your decision will likely depend on the given measures in the problem. The

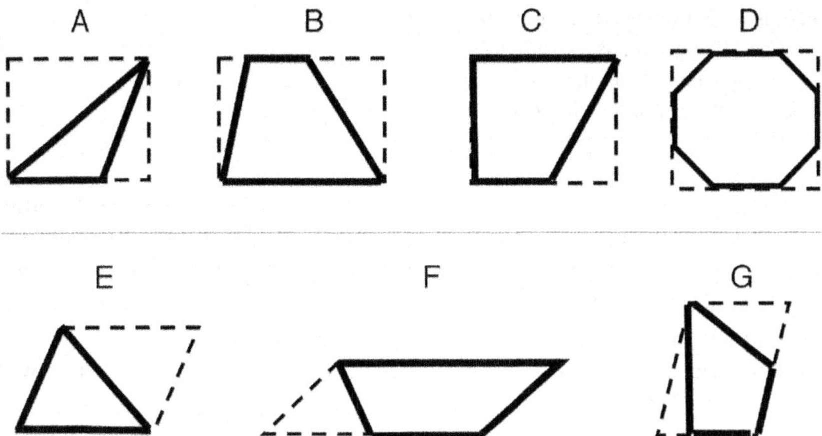

Figure 5.8 Unfamiliar shape areas can be surrounded by rectangles or parallelograms, then remove the extra pieces.

information provided in the problem will depend on the difficulty level and how much problem-solving effort is required. For example, a diagram like "A" in Figure 5.8 for a problem could state that the area of the solid-lined triangle is nine square units and that one side of the dashed-line square is six units.

One question for that problem could be, "What is the combined area of the two extra triangles?" To answer this question, a student would need to calculate the area of the square as thirty-six square units and then subtract the nine square units in the solid-line triangle. The leftover total for the two extra triangles would be twenty-seven square units. Just as Theo's test question did not require the measure of the separate angles, neither does this question request separate triangle areas. The question asks for the total of both leftover triangles.

Another question for a problem with this diagram with the same information provided could be, "What is the area of just the lower right small triangle in diagram A?" To answer the second question, a student would calculate the area of the square for the thirty-six square units and then subtract half of that (eighteen square units) because one of the extra dotted triangles is half of the square. Next, they would subtract the nine square units belonging to the solid-line triangle and end with nine square units in the small lower-right triangle. There are other ways to answer these two questions. Use your imagination to try to find new questions and new solutions.

The rectangle area (base times height or length times width) formula also works with shapes if your child is brave and flexible enough to try different arrangements by cutting and rotating pieces to make it fit into a familiar shape. For example, Rita wrote a poem about how the parallelogram in Figure 5.9 could be decomposed and the triangle slid over to make the familiar rectangle. The original parallelogram and the new rectangle have the same base (length), same height (width), and therefore the same area.

Area of a rectangle is an easy affair
All you have to do is count up the squares.
Don't stop now, there's another way

Area equals length times width, and I've got more to say.
Area of a parallelogram has its own groove.
Borrow a right triangle and make the move.
Now you've got a rectangle, and it's easy to see
Base times height is what it will be.

The decomposition and rearrangement will also work with a trapezoid (Figure 5.9). The right triangle is cut away, moved to the other side, and turned upside down to make a rectangle. The trapezoid and the area have simply rearranged pieces so that it will still have the same area. The idea is to be flexible in how you see decompositions and use them to your advantage. As long as you keep the measurements intact, you can move around decompositions of a shape and still have the same area of your new shape.

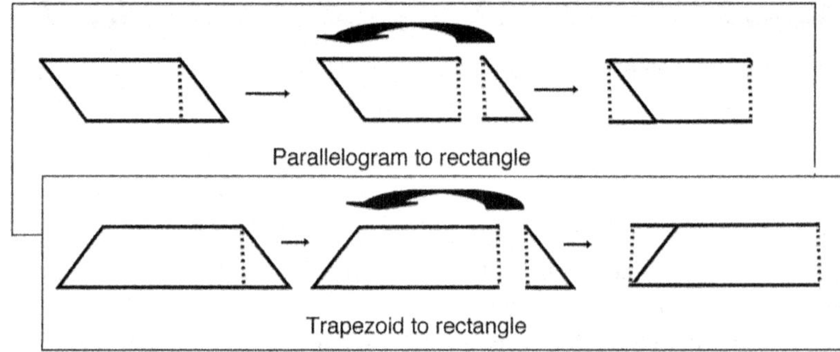

Figure 5.9   Parallelogram rearranged to make a rectangle; trapezoid rearranged to make a rectangle.

Sally, a fourth-grade student, wrote a description of her interpretation for calculating the area of a circle after exploring how to cover a circle with right triangles. Sally rotated the triangles about the center of the circle, sketched different arrangements, and typed her explanation, shown in Figure 5.10. Even though she knew that squares would cover the circle area, she chose to use these right triangles instead, reasoning that triangles would cover more of a circle with fewer leftovers than a square.

Sally's description: "To find the area of a circle, first I thought of how to get the area of any shape. First you measure the height and multiply it by the length and you get the area. But since a circle has no sides, I decided to give it some. I also decided to split it into 16 parts. . . . Then I saw that I had made 16 triangles. So I knew that if I found the area of one of the triangles and multiplied it by sixteen I will get approximately the area of the circle."

As long as we are decomposing polygons into triangles, the apothems can come in handy in a strategy to calculate the areas of the polygons. Remember, these apothems are the heights of all of those triangles from Chapter 4. The apothems (heights) are multiplied by each respective triangles' base to calculate the triangle areas. As one might expect, there's a formula to calculate the area of a regular polygon that includes one apothem's triangle height (the same apothem fits all of the triangles). That formula is explained, illustrated, and *decomposed* into triangles in Chapter 11 and reflects the spatial sense in Chapter 1.

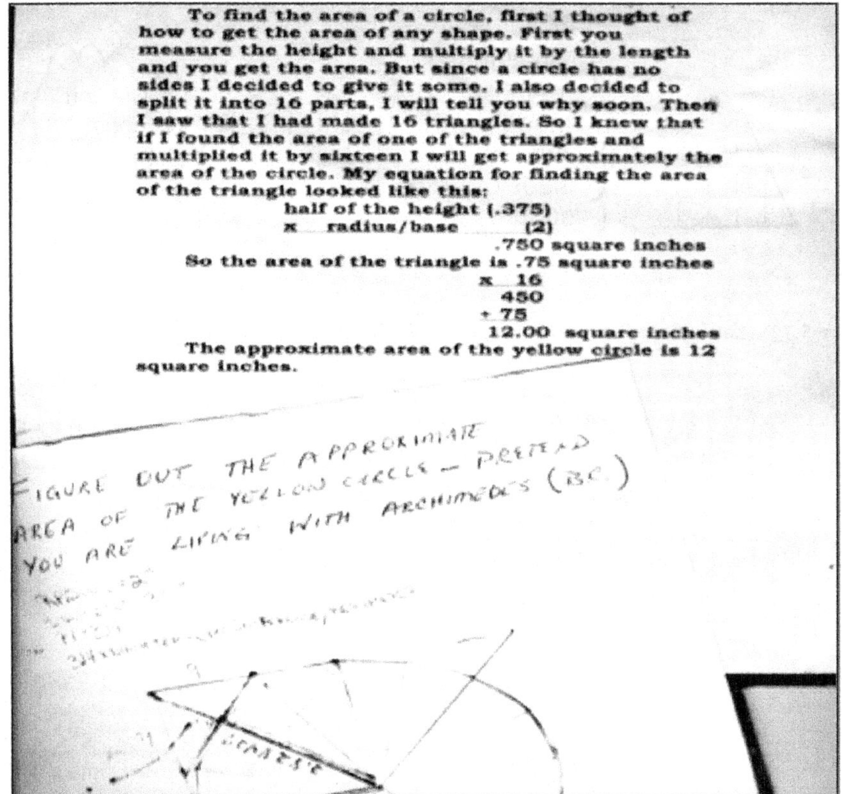

Figure 5.10   Sally's exploration with circle area.

KEEP IN MIND

Areas of polygons lose their mystery when they can be associated with more familiar and reliable shapes. They can be decomposed and added, or encased in a more familiar shape so that the extras can be subtracted and cut off, or simply rearranged into a familiar shape. Formulas are also available. However, if the letters and purposes get "mushed" (Katrina's words) like they did with Katrina, the formulas can hurt more than help.

How Katrina Clarified Her Confusion of Area and Perimeter Formulas

Unless your child has a good handle on the difference between area and perimeter with an understanding deeper than just the formula, perimeters of triangles and any other polygon that does not have corner squares can bring confusing challenges. Perimeter can also be tricky because it too often shares the same picture with area. If your child has seen the same picture for two different concepts, that can be very confusing and even feel duplicitous to your visual-spatial child. "Duplicity breeds distrust."[4] Also, as shown in Figure 5.11, counting edges of a triangle is not exactly clear because the triangle's edges *are not the sides of squares.*

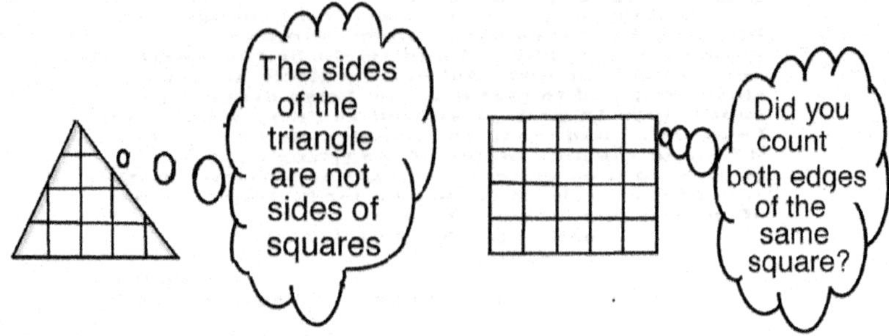

Figure 5.11   Katrina's problem with triangles showing only parts of squares in their area.

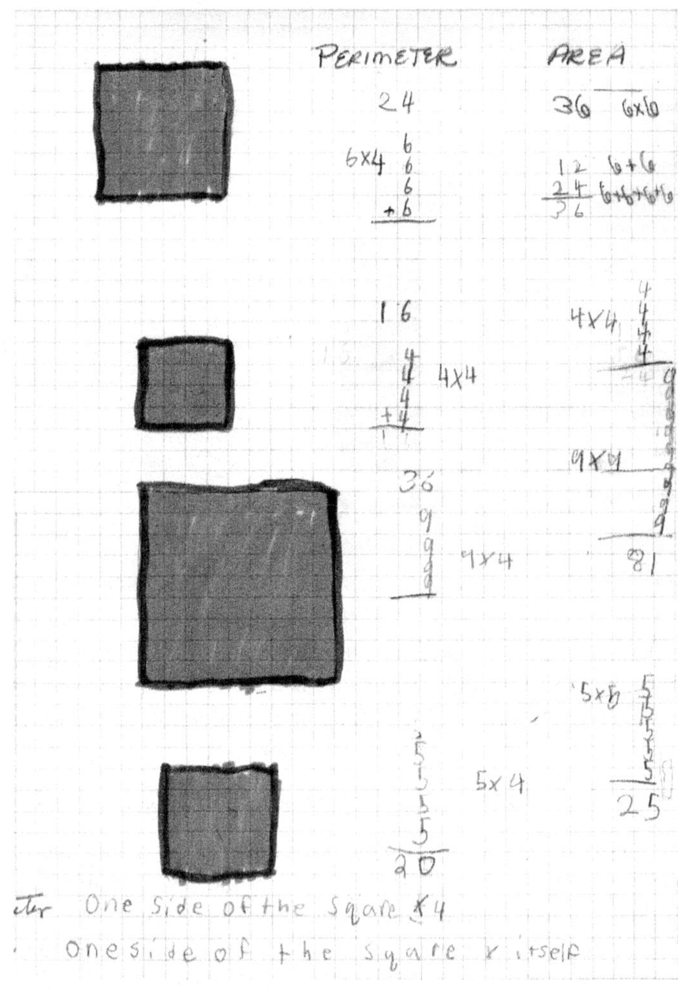

Figure 5.12   Katrina's solution to clarify area and perimeter.

Even though Katrina could describe the difference in the concepts, she nevertheless always used the four sides as a multiplier for both concepts instead of distinguishing the *coverage for area* as separate from the *edge meaning of perimeter.* After coloring, counting, and writing the computations in the two columns (see Figure 5.12), Katrina's words for perimeter were, "one of the sides times four" and for area, "one side of the square times itself." One might think that this had already been explained to her this way; however, she needed to make her own drawings and collect her own data in order to make her own personal sense of these concepts.

When asked to draw a picture of perimeter and a *different picture* for area, Katrina thought awhile, made a few attempts, and then she chose to draw the squares in Figure 5.12 (the squares *are different* according to Katrina). Katrina colored the area red and the perimeter blue, counted the units, noted the calculations, and then wrote the words *Perimeter* and *Area* at the top of the two columns. Although the "four" from the number of sides was in every *perimeter* calculation of the squares, both multiplication numbers changed for calculating the square *areas.*

One of Katrina's strengths as evaluated in her Individualized Education Plan (IEP) was recognizing similarities. Whether that similarity strength influenced her choice of using all squares instead of other shapes is not clear. This choice made sense to Katrina because her assessment two weeks later indicated that the interpretations were no longer "mushed" and, yes, she was able to generalize her new learning to distinguishing area and perimeter of other shapes. Allowing Katrina to draw her own pictures and interpret her own meanings, *even though the pictures may look the same to others,* certainly made the staying-power difference for Katrina.

NOTES

1. Abbott, *Flatland*, 4.
2. Boyer, *A History of Mathematics*, 180.
3. Roy, *The Enigma of Creation and Destruction*, 27.
4. Covey, *The Seven Habits of Highly Effective People*, 21.

Chapter 6

Stack and Pour to Fill Volume in Spaceland

"I am indeed, in a certain sense a Circle," replied the Voice, "and a more perfect Circle than any in Flatland; but to speak more accurately, I am many Circles in one."[1]

—Description of a Spacelander, an inhabitant of Spaceland

We live in a three-dimensional world like the Spacelanders, a world that extends "up above and down below" from those populated by the two-dimensional shape inhabitants in Flatland. That simplistic, yet powerful, expression by the Voice can help to visualize the properties and characteristics of Flatlander polygons as being different from the polyhedra that live in the horizontal *combined with the vertical* territory of the third dimension. This chapter is about how those three-dimensional properties of Spaceland can influence us, how they make patterns, and how they help us make sense of the world around us.

In our three-dimensional world, we have organized our number system into place value categories that reflect the different dimensions. Those first four categories from place value are ones, tens, hundreds, and thousands. The thousands category has taken our place value for numbers into the third dimension by multiplying length, width, and height and is expressed as "10^3" or "ten cubed." The height for this *cube* lives in the "up above" location extended from the flat *square* expressed as "10^2," an inhabitant of Flatland, and the "10^2" Flatlander was extended from the "10^1," who lives in Lineland.

Just as the zero-dimensional point assembled into the one-dimensional line, the one-dimensional linear shapes generated two-dimensional shapes, such as squares, triangles, and other polygons. The building of dimensions continues so that the two-dimensional shapes build three-dimensional shapes. These three-dimensional shapes become part of the "up above and down below" when they are folded from two-dimensional shapes (known as "nets") and are composed of flat surfaces called "faces."

After these nets are folded into new three-dimensional shapes, then the new shape has edges and vertices where the surface faces meet. When all of these faces are painted, or when the areas of the faces are added together, then they are known collectively as surface area. If and when the polyhedra is filled up or stacked with lots of levels, then that total internal measurement is assigned the name "volume."

POLYHEDRA NETS MAKE THREE DIMENSIONS
FROM TWO DIMENSIONS

The two-dimensional shapes are the faces of a three-dimensional solid. One way to "see" how the assembly of two-dimensional faces make three-dimensional shapes is through nets. Nets decompose the three-dimensional shape by cutting it apart and laying it flat. Two examples of nets are shown in Figure 6.1 for a cylinder (tube) and a rectangular prism (or shoebox). Remember, only the flat-sided shapes are called polyhedra. The cylinder and sphere do not have flat sides, so they keep their own names and are recognized as solids.

Before you toss that cereal box into the recycle bin, cut out all of the sides of the box and trim the flaps. The front, back, two sides, bottom, and top of the box is what would be left after your cut-away pieces. The result is a homemade rectangular prism net. Retrieve a cylinder (preferably cardboard) package for chips or cleansers and cut the long side of the tube and spread it open to see the flat rectangle that was before it was folded into a cylinder tube. The round top and bottom are the circles that are part of the cylinder net. Search around the house and find some other nets to cut and unfold.

Nets provide a different way to "see" three-dimensional shapes so that you can select which part of the shape you will need to solve a particular question or problem. For example, nets are the cut patterns for manufacturing product packages like the ones shown in Figure 6.1. Someone needs to design the product and make decisions about how to cover the outside surface (surface area). Someone else must figure out how to cut out the container package (polyhedra net) in order to manufacture it for the customer.

Like Plato, Archimedes, and other mathematicians before (and after) them, your child can use shapes and manipulatives like Magformers and Polydrons to explore the five Platonic Solids and the thirteen Archimedean Solids, some of which are shown in

Figure 6.1 Three-dimensional nets.

Figure 6.2. There are more three-dimensional polyhedra shapes at the ready for you and your child to find and make, like the twentieth-century ninety-two Johnson Solids and the nine Deltahedrons, plus others.[2]

Platonic Solids are more frequently found in elementary geometry studies than are Archimedean Solids or the others, but that should not stop you or your child from building them by making nets and folding them into polyhedrons—we are long past the dinosaur era, but that doesn't stop children's interest in dinosaurs. Making them takes a bit of dexterity and patience, but is well worth the effort in order to develop spatial sense. These photos in Figure 6.2 show all five Platonic Solids and four of the Archimedean Solids.

Five Platonic Solids:
L to R bottom: tetrahedron,
cube, octahedron. L to R top:
dodecahedron, icosahedron

Four Archimedian Solids: L to R
bottom: truncated octahedron,
truncated cube. L to R top:
cuboctahedron, truncated
tetrahedron.

Figure 6.2   Platonic and Archimedian polyhedra solids.

SURFACE AREAS, POLYHEDRA NETS, AND ORANGE PEELS

There is an advantage to knowing about polyhedra nets when learning about surface area, primarily *because* folding a net of two-dimensional shape areas makes a three-dimensional solid. Surface area of solids as a topic is usually reserved for upper elementary school or middle school, and it is always presented in a high-school geometry class. However, young children can informally learn about the idea of surface areas by building their own solids.

Seventh-grader Minton carefully peeled an orange, tore the skin into small, almost flat sections, and then placed them on four circles, doing his best to cover all of the space inside the circles. These four circles had the same diameter as the orange, and actually came from the cross-sectional slice through the middle of the orange, like the one in Figure 6.3. The topic in his seventh-grade math class included the formulas for finding surface areas, but Minton had difficulty understanding how the sphere could have a surface area. The polyhedra had flat surface faces, but the sphere was different. Seeing how the orange peel pieces covered the circles helped Minton understand.

Exploration with 3-D activities builds spatial thinking. Friedrich Fröebel's Gifts and other building sets provide a wide range of learning experiences, especially

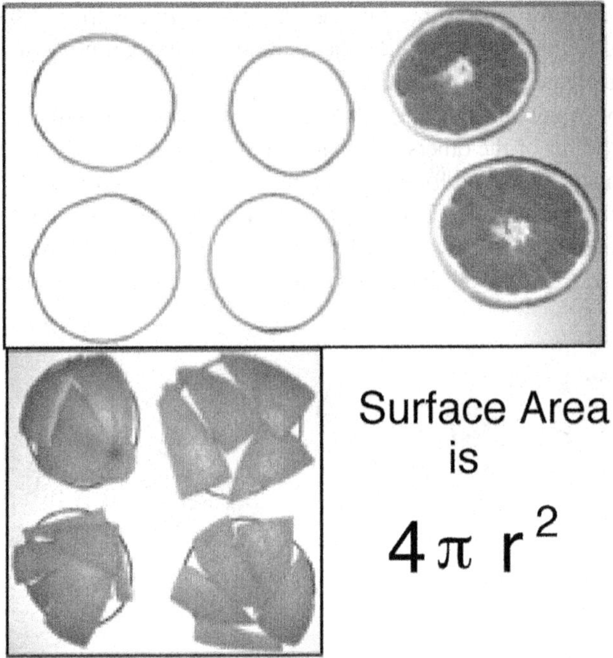

Figure 6.3   Surface area of sphere with orange peels.

for early learners. Older children, who may be budding architects like Frank Lloyd Wright, young inventors like Buckminster Fuller, topology thinkers like Abigail Thompson, or Nobel Prize winners in mathematics like Maryam Mirzakhani, will also benefit from building experiences. No one is ever too old to develop spatial abilities! "The seemingly simple activity of manipulating building blocks requires an amazing number of complex thinking and academic skills."[3]

FILL 'ER UP FOR VOLUME

Perimeters outline, areas cover, and volumes fill. Sounds simple enough, so why does the idea of calculating the volume of something make many children—and adults— cringe? Volume is very often presented as a formula rather than a filling or stacking activity. The volume formula, like its area and perimeter predecessors, is not the volume any more than that map is the territory. A volume is *calculated* using a formula or with one or more of the action verbs of addition, and multiplication. Most people use multiplication because that was how they learned to compute volume. Addition works too when the different layers in the container are added.

When you pour, you fill; when you stack, you fill. Both pouring and stacking are action verbs and serve as volume-finding activities. If you have transparent, plastic, see-through containers, then pouring water mixed with food coloring helps your child monitor the amount as the container fills. This is when "actions can speak louder than words." By giving your child the opportunity to pour water or stack layers, they can "see" the interior, and not just take it on faith that the picture of a container or a closed

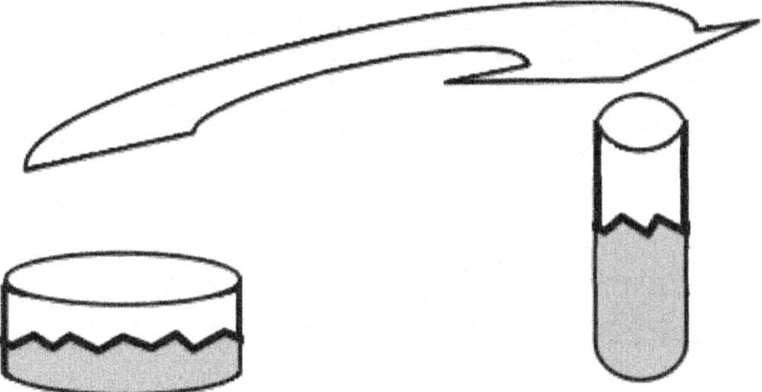

Figure 6.4 Piaget's pouring illustration.

or solid box has something inside of it. Taking "on faith" that an interior has contents doesn't build spatial sense. "Seeing is believing" works better.

Pouring into different containers with that color infused water liquid will help your child get a better spatial sense of volume. Your child might judge the amount of liquid (even if they watch the pouring happen) based on the size of the jar rather than the same amount poured contents. Your child will say that the amounts in the two containers in Figure 6.4 are different and the tall container has more water if they are still in Piaget's preoperation developmental stage, when children do not yet recognize conservation of space. They believe that just because the liquid level is higher then there is more liquid, ignoring the change of container shapes.

A letter like "h" can represent "height" for the number of layers in the volume formula since water doesn't appear to have layers. Just like the area relationship between a rectangle and a parallelogram, both depend on heights (although the rectangle and parallelogram sometimes use length instead of height). The volume of any polyhedra or cylinder, even if slipped sideways like the examples in Figure 6.5, will have the same volume if the height and base of the polyhedra remain the same, just as a two-dimensional parallelogram and a rectangle can share the same area formula as long as *the height measure and the base measure are the same in both shapes.*

Because liquid is not as rigid as a disk, the volume relationship illustrated by stacking may not be as obvious to either you or your child. Seeing the "inside" of a cube when the cube, or other solid, is opaque can present a challenge to anyone who has not been able to imagine a cross section or otherwise *recognize what is inside* the solid. You will know that the "inside" is missing in their mind when your child tells you to add all of the sides (faces) to get the volume. They say that because that is *what they can see.*

Mica's assignment was to count cube faces made with different numbers of cubes and complete a table of numbers of painted faces. Part of his table is shown in Figure 6.7. Each cube size (1x1x1, 2x2x2, 3x3x3, and so on) was painted only on the surface so that the cubes on the exterior had at least one painted face and none of the interior cubes had any paint. While he worked with the different cubes, taking them apart to count the painted faces and the nonpainted cubes, it became clear that when the cubes were reassembled, Mica no longer "saw" the interior cubes. Mica operated on the idea of "what you can see is what you can use."

As the sample 3x3 and 5x5 blocks show in Figure 6.6, the interior blocks in the 5x5 cube have a different color from the external cubes so that Mica could see the

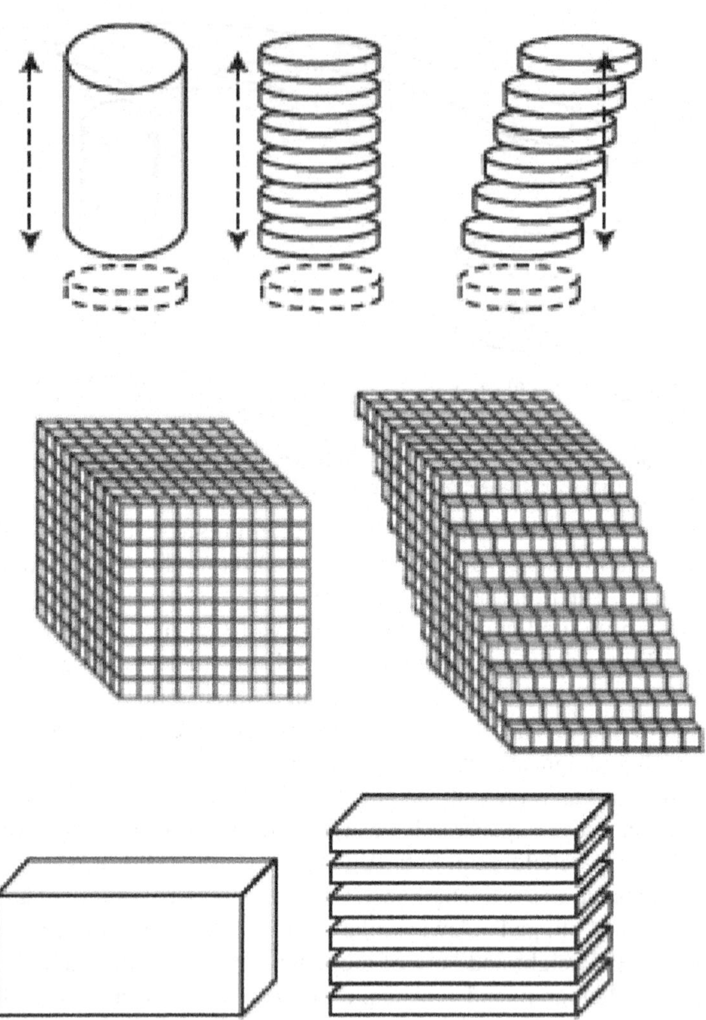

Figure 6.5   Three-dimensional "stacking" for volume.

difference and "see" that there were indeed cubes in the middle that still needed to be included for volume even when the cube was reassembled. He was one of the children who relied on symbols in a formula, not the meaning. The 3x3 cube also shows different colors for interior and exterior cubes so that Mica could see that it, too, had interior cubes to count. The painted cube pattern lesson turned into a volume lesson before it returned to the pattern sequence focus.

The original painted-faces lesson was about surface area, patterns of painted faces, and how these numbers were connected to the math topic for finding the number of faces, edges, and vertices of cubes (or eventually to other polyhedra like those Platonic and Archimedian solid polyhedra). The first task was to count the total number of cubes in the 3x3x3 cube, then count the number of individual cubes that had (1) one face painted, (2) two faces painted, (3) three faces painted, and (4) no painted faces. A section of Mica's record is shown in Figure 6.7, showing the totals as he increased the number of cubes to 4x4x4 and 5x5x5 and so on.

Closed 5 x 5 x 5 cube

Opened cross-section for
5 x 5 x 5 cube

Closed 3 x 3 x 3
painted surface cube

Opened cross-section
for 3 x 3 x 3 cube

Figure 6.6   Mica's cubes: 3x3 and 5x5.

Number of painted faces

6	5	4	3	2	1	0	Tot	Eq	
0	0	0	8	12	6	1	27	3³	3x3x3
	0	0	8	24 (12×2)	24 (4×6)	8	64	4³	4x4x4
0	0	0	8	36 (12×3)	54 (9×6)	27	125	5³	5x5x5
		0	8	48 (12×4)	96 (16×6)	64 (4³)	216	6³	6x6x6
		0	8	60 (12×5)	150 (25×6)	125 (5×25)	343		7×7×7
0	0	0	8	72 (12×6)	216 (36×6)	216	512		8+8×8
0	0	0	8	84	294	343	729		9+9×9
0	0	0	8	(7×6)					

Figure 6.7   Mica's record sheet.

Mica found his own pattern for calculating the number of cubes. It was not immediately apparent that his description of his pattern would generalize to other cubes. However, because he used selections from the previous number of cubes in his record, his calculations did prove to be consistently accurate. One of the goals for the activity was to show how each newly constructed cube contained a previous cube (and the volume as Mica also learned). Mica recognized this relationship with numbers first, before he recognized it as connected to volume. After the brief volume interim lesson, Mica connected the two concepts *and the association with exponents.*

Your child might also enjoy a box-filling activity that has made the rounds in books and activity programs adapted for middle-school, high-school, and calculus classes over the past several decades. It is easy to do and may be helpful to your child as an activity at any age, depending on their level and interest, and it can bridge the volume ideas of pouring, filling, and stacking to fill the "inside" of a container. All you need is a lot of centimeter grid sheets from office supply or educational stores

---

Try This:

Start with several sheets marked with centimeter squares (like graph paper). The sheets have the same dimensions length and width. Inch-squared papers and cubes will also work. Cut out the corners of each sheet and fold up the sides to make a box. Increase the size of the corner each time, until you can no longer make a box.

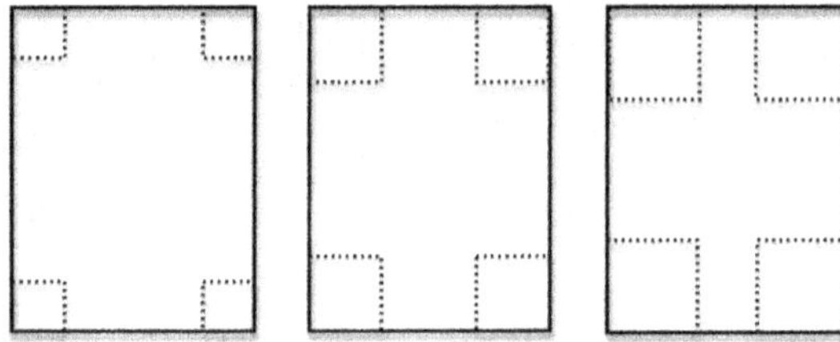

Cut out the corners and make boxes to fill.

Fold up each box, calculate the volume, write the volume on a record sheet, keeping track of all the measurements.
Compare the measurements.

Which box has the greatest volume?

Ask some questions of your own about the data in your record sheet.

Figure 6.8 Task Box: Try This.

(or downloaded) and centimeter squares (or another grid with matching-measure squares, such as inch cubes and inch squares) to fill the cut-out boxes.

Filling a box has its advantages because children like Mica can see the volume increase as the bottom layers stack up. They can see it happen *as it happens* instead of just seeing the final picture of a box or other container. Jo was filling her boxes and estimating the bottom layer amounts as she cut out the corners and folded up the sides. As in Mica's case, there was something awry in her estimations. Jo's estimates for the base coverage were very low, too low for the layers. Although the lesson was on volume, another opportunity about estimation and multiplication presented itself for exploration.

Jo had worked with base ten blocks, so she was familiar with the concept of area as a spatial arrangement with the blocks. She knew that if a hundreds block could fit on the bottom layer, then the layer has at least one hundred units. Her base-ten multiplication experiences had ended with the two-dimensional areas. The base-ten multiplication arrangements in Figure 6.9 show three of the first layers for three of the boxes she built for volume. Now as she multiplied these area layers by the *height stack of layers,* the volume amounts made more sense.

This kind of box building and filling can help children across several grade levels to get a feel for volume concept. Younger children will enjoy making and comparing boxes and even filling them; older children will be able to fill the boxes and compare volume amounts of the different sized boxes. If your child has collected data or kept a record of his work, then they can fill these boxes and graph the data. When they graph the data, the graph will show how the volume amounts change as the corner sides of cut outs increase. The photo in Figure 6.10 shows how the boxes were physically attached to their related graphed data to enhance the visual idea.

Figure 6.9   Measuring different volume bases to calculate different volumes in boxes.

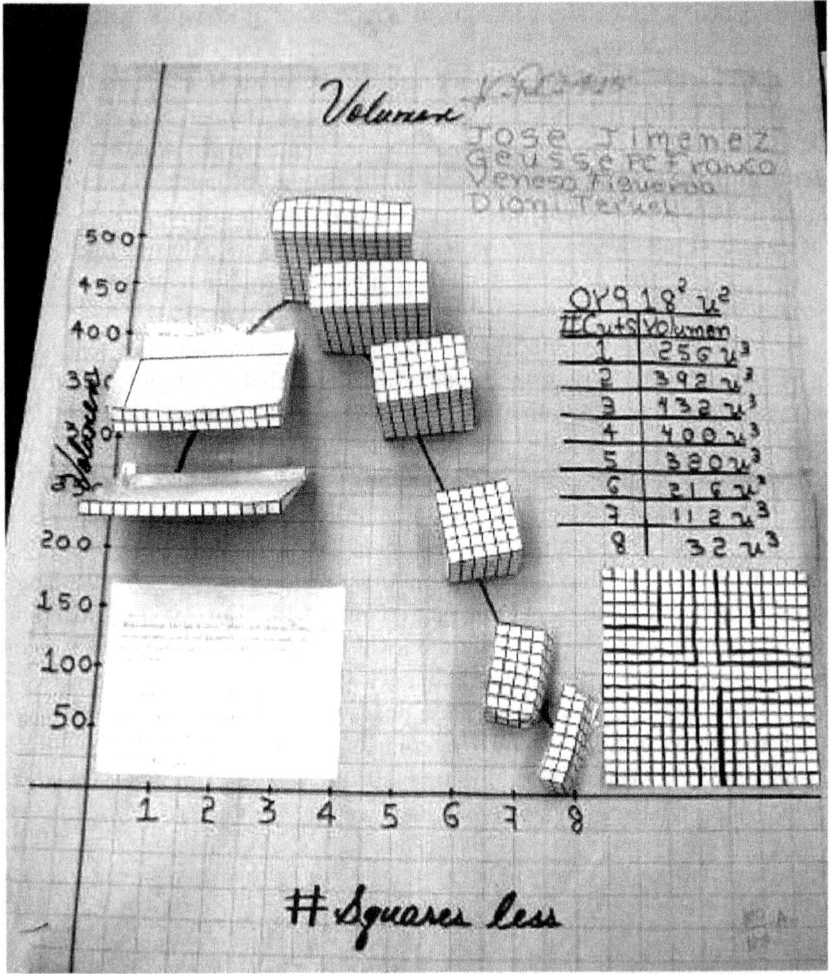

Figure 6.10   Graph of box volume experiment.

KEEP IN MIND

Calculating volume probably begins around the fifth grade; however, the three-dimensional exposure can begin as early as preschool with those building blocks. As children learn to calculate with different numbers, it is very helpful to them to know that all of these new topics are connected. The "3^3" read as "three cubed" as an exponent, multiplication with three numbers, filling containers for volume, and the 1,000 block in place value are all connected ideas.

Geometry isn't a finished study and it isn't limited to a fixed set of formulas, but it *is* organized. There is still plenty of space left for exploration. The only thing that limits your child is imagination. If your child stays within Euclidian space, then they must abide by the boundaries determined by Euclid. Of course, your child, like Riemann or Lobachevski, is invited to venture out into uncharted thinking but, also like Euclid (flat plane) and Lobachevski (hyperbolic plane) or Riemann (curved plane), they will need to develop her own new set of definitions and properties and define her own boundaries.

> Even the mathematicians would like to nibble at the forbidden fruit, to glimpse what it would be like to . . . to slip a moment into a fourth dimension.[4]

—Kasner and Newman

NOTES

1. Abbott, *Flatland*, 68.
2. Johnson, *Building Geometry*, 6.
3. Carter, "Move Over Frank Lloyd Wright," 8.
4. Kasner and Newman, *Mathematics and the Imagination*, 124.

Part III

SHAPE RELATIONSHIPS AND PATTERNS

Nature is relationships in space.Geometry defines relationships in space.Art creates relationships in space.[1]

—Newman and Boles

Are spatial relationships and imagination really as entwined in mathematics as they are in art and nature? The answer to this question might depend on who you ask: Escher, the artist; Mandelbrot, the mathematician; or Kepler, the scientist. The extent of the relationships and patterns in geometry is as big and as expansive as your imagination or your child's imagination will allow—as large as nature in her arrangements of space and as diverse as art in the creations within space. Nature, geometry, and art clearly have their own separate specialties and gifts, but they most certainly coexist as experiences in mathematical thought.

Being able to recognize these entwined relationships will depend on your child's willingness, and yours too, to explore, guess, estimate, have some wild thoughts for conjectures, test, confirm truth, and reach conclusions useful to others. George Pòlya encouraged mathematical guessing, strategic guessing, and evaluating through tests, while Paul Halmos viewed mathematics as a creative artful endeavor that supports trial and error, experiment and guesswork. Earlier mathematicians like Bombeli had wild thoughts about imaginary numbers and Nicholas Chouquet considered negative integers "absurd."

M. C. Escher's woodcuts that play with dimensions and the infinity concept are beautiful examples of artistic imagination. Benoit Mandelbrot imagined fractals during the twentieth century based on concepts initially explored in the seventeenth century. Johannes Kepler improved on Copernicus' circular planetary orbital work by using his imagination, and redefined those orbits as elliptical paths. Each artist, mathematician, and scientist "stood on the shoulders of [previous] giants," a quote generally attributed to Isaac Newton even though he no doubt was following the lead of the twelfth-century philosopher Bernard of Chartres.

Besides the more familiar interconnections between geometry and nature, like the regular pentagonal shape of the starfish, the bee's hexagonal honeycomb, the pattern sequence in the arrangement of flower petals, and the Golden Ratio visible within a nautilus shell's spiral, there are always new places for the imagination to stretch a bit further. All we need to do is look closer at the geometrical designs within art and the geometric relationships in nature.

This section examines a few of the many geometry relationships and patterns. Chapter 7 explores similarity and congruence. Chapter 8 describes several unique comparisons, some having direct associations to special shapes. Chapter 9 helps to get your child's imagination in high gear by suggesting several unlikely, yet predicable, ways when shapes change. These relationships and patterns are threaded and woven into your child's arithmetic and geometry curriculum over many grades.

Just by exploring and examining these relationships, you and your child will be like mathematicians throughout history—Bombelli, Chouquet, Pólya, and Halmos. By investigating special patterns and unique relationships of shapes by using your best of Pólya, Halmos, Poincaré, or Bombelli-type guesses, your child will learn how to identify the dependable kind of predictable patterns and relationships, *and* be able to remember them. Who knows? Your child may even discover some patterns and relationships of their own!

NOTE

1. Newman and Boles, *Universal Patterns*, xvi.

Chapter 7

Similarity and Congruence

The two fundamental categories [number and shape] are brought together
through the unifying concept of measurement.[1]

—Peter Hilton and Jean Pedersen

The more flexibility in thinking that your child can develop, the more comfortable
they will be as they predict, describe, and measure congruence in different arrangements *and* identify similarity with proportional measurements in different circum-
stances. Measurements can unify geometry concepts and geometric diagrams can
clarify measured relationships. Symmetry, one of those spatial concepts, requires
congruence because symmetry requires *an exact match* between shapes, whether as
an internal or an external arrangement.

Just as the square is a specific kind of rectangle and a fraction is a specific kind of
ratio, congruence is a specific kind of similarity. Similarity can involve a change by
enlargements or reductions of shapes; congruence cannot. Specifically, a similarity
relationship between shapes requires that the angles remain congruent while the sides
can shrink or grow in proportion to each other. The congruence relationship, on the
other hand, requires that the angles and sides of congruent shapes stay the same. Congruence (the symbol is ≅) is the specific kind of similarity (≈) when the proportion is
one-to-one. Interestingly, the symbols are related, too.

SIMILARITY ALLOWS STRETCHING AND SHRINKING

Magnifying glasses and microscopes are all about similarity, enlarging and reducing
images. These two instruments show very large and very small versions of the same
image—proportional images. "Similar" as a description is used in the English language to describe a likeness but not necessarily an identical relationship between two
things. In math, the usage has the same notion because similar shapes *must have the
exact same shape* yet can have proportional sides. The examples in Figure 7.1 show

Similar Shapes

Figure 7.1   Similar shapes.

what similarity can look like in shapes. Showing shapes as "nested" or overlapping makes it easier to see how angles are kept the same.

Trace the shapes and move them within each nest to see how the angles match up and align with all of the angles. One side of a similar shape cannot shrink (or grow) out of proportion to another side. Just as the equivalent fractions must have the same rate of change in the numerator and denominator, so do similar shapes need to keep the same rate of change for all of their sides. If the same rate of change is not used, then the shape would look like the distorted image in that carnival mirror. Your child can recognize these similar shapes in architecture designs, spider webs, needlepoint, quilts, rugs, and tree rings (assuming same annual weather).

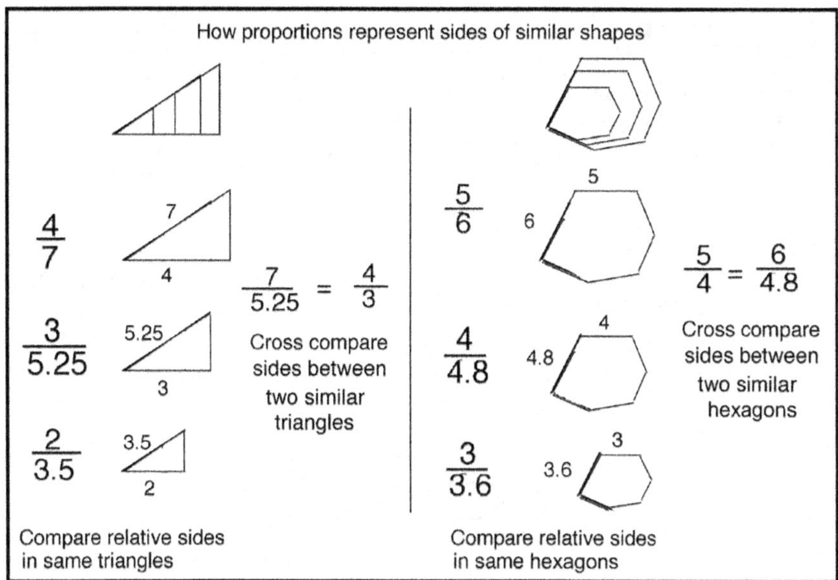

Figure 7.2 Proportional ratios.

What happens to one side of a shape also happens to the same relative side of its similar shape partner. *Proportional ratios* that compare two sides in the same shape will also apply to two relative sides of a similar shape. Just as a magnifying glass or a telescope enlarges all parts (except the angles) of the item seen "through the looking glass," the multiplication of side measures works the same way *on the respective sides*. Use your calculator to confirm your own truth for these proportions shown in Figure 7.2. Notice how the sides are compared using proportions within the same shape and between shapes.

Are all pentagons similar to each other? How about all rectangles or triangles? No, not unless the pentagons are regular pentagons, the rectangles are regular rectangles (also known as squares) and the triangles are all regular (also known as equilateral and equiangular). Does it look like all circles will be similar? Do they all have 360°? Do you think your child could draw two different circles that would *not* be similar? Are all ellipses similar? What is it about circles that will guarantee that they are all similar to each other, regardless of their size? What is it about elliptical shapes that make their similarity relationship dependent on their measurements?

Many questions spark lots of spatial-sense thinking. The ellipse has rounded edges just as a circle's boundaries (circumference) are round. Does pi apply to ellipses, too? For now, *yes* answers the question, "does the pi ratio apply to both circle and ellipse?" and *yes* to the question, "are all circles similar?" *No* answers the question, "are all ellipses similar?" because an ellipse has two different-sized line segments that go through its center, unlike the circle's diameter, which are the same length all the way around. Chapter 8 offers much more information about pi and its uniqueness related to rounded shapes.

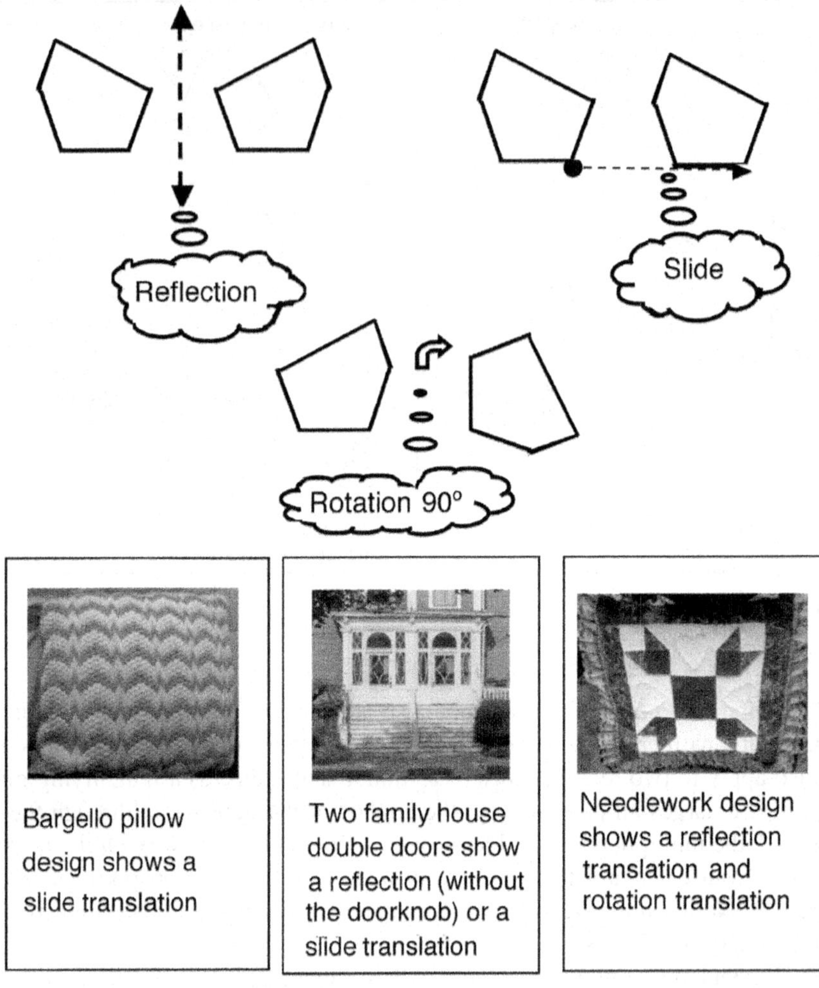

Figure 7.3 Transformations: reflection, rotation, and slide.

CONGRUENCE, TRANSFORMATIONS, AND TESSELLATIONS

Congruence is a special category of similarity in that congruence is the one-to-one proportional relationship of similar shapes. This concept can appear in your child's homework, in classwork, or in shapes anywhere around them. When your child can recognize the consistency and even replication of the same shape and size, then many of the newer ideas and concepts become easier to learn and remember. There is something comforting about congruent dependability of size and shape. Congruency appears in different spatial arrangements and settings, so the more experiences your child has with recognizing congruent shapes, the better.

Transformations give permission to move things around, both literally with physical items and mentally with imagination. These transformations have nicknames as "flips, slides, and turns," (Figure 7.3) but the more formal terminology is reflections,

Figure 7.4 Task Box: Ask Yourself.

slide translations, and rotations. All transformations keep the flipped, turned, and slid shape intact and congruent. A reflection depends on a line of reflection (the one in Figure 7.3 is an external line of symmetry); an angle measurement is required for a rotation, and a direction is needed for a slide translation transformation. The spatial sense in transformations is about flexibility and keeping congruence.

Examples of these transformations are all around us in architecture, needlework, furniture design, baskets, rugs, beehives, seashells, hubcaps, dance choreography, crystals, and many other places and objects surrounding us—too many to list them all. They are just waiting for you to "see" them—or create them yourself. Use a little bit of ingenuity, a lot of imagination, some tracing paper, paint, folding paper, and any drawing tools you can find to help you create your own designs.

A design is called a tessellation if there are no gaps between the geometric pieces when tessellated items are assembled *and* the total degree measure around all vertices where they meet totals 360°, the same number of degrees in a circle. In Figure 7.5, there is a dotted circle where the corners (vertices) meet for the regular hexagon tessellation and the regular (equilateral) triangle tessellation. Usually by the fourth and fifth grades, your child will cover areas of surfaces with tiles and tessellate with shapes. There are some slight spaces between the triangle pieces in Figure 7.5 to show how the polygons were rotated, some reflected, and all slid together.

As Sara and Margo found out, tessellations are areas artfully put together. Sara created her own shape to tessellate, cut it out, and then reassembled it. Margo not only created an unusual mushroom shape to tessellate, she decorated the pieces with *Peanuts* characters so that when her friends reassembled the puzzle, they had to match the shape of the mushroom *and* the *Peanuts* character. Besides the benefits of using imagination and creativity, the math content for tessellations includes perimeters,

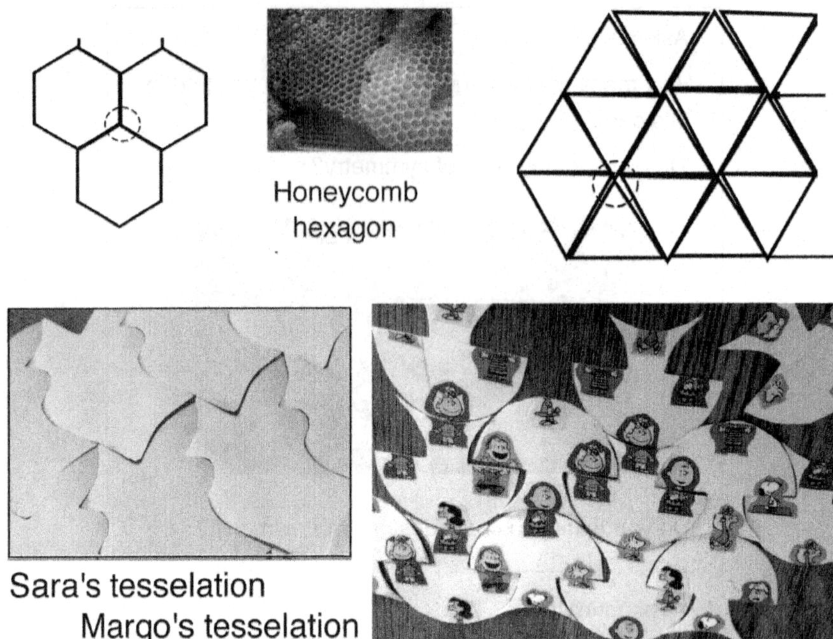

Figure 7.5   Sara's and Margo's tesselations and hexagon honeycomb tesselation.

areas, angle measures, total circle degree measures, length measures, congruence, decomposition and composition of shapes, just to name a few topics.

The twentieth-century artist M. C. Escher would use tessellations as part of his artistic woodcuts. He would select a polygon that would tessellate, create a design inside that polygon, and then replicate the design several times to create an amazing design. His art was very mathematical. Imagine the mathematics that it took to create his designs, as well as his incredible artistic talent. Sara, Margo, and Escher understood the mathematical relationships behind tessellations from the inside out. As long as equilateral triangles can decompose polygons, these same polygons can create tessellated art.

The regular pentagon shape itself does not tessellate (see Figure 7.6) because the sum of the angles where the pentagon meet will not complete the necessary 360° circle. Three regular pentagons can be arranged with sides touching, but the total angle is only 324° where the pentagon corners meet. Each angle of the regular pentagon measures 108° and 108°x3=324°. Chapter 11 includes more information about how these angles are calculated and how regular pentagons and other regular polygons subdivide into congruent isosceles triangles.

The regular pentagon may not tessellate, but it contains several special comparisons and ratios that continue to fascinate mathematicians. One example of a special ratio is found in a five-pointed star (like the starfish) and in a specific rectangle found in the nautilus shell (and in Greek architecture) shown in Figure 7.6. The special comparison found in the regular pentagon, the nautilus shell, and starfish, along with other special number comparisons, are explained and explored in Chapter 8.

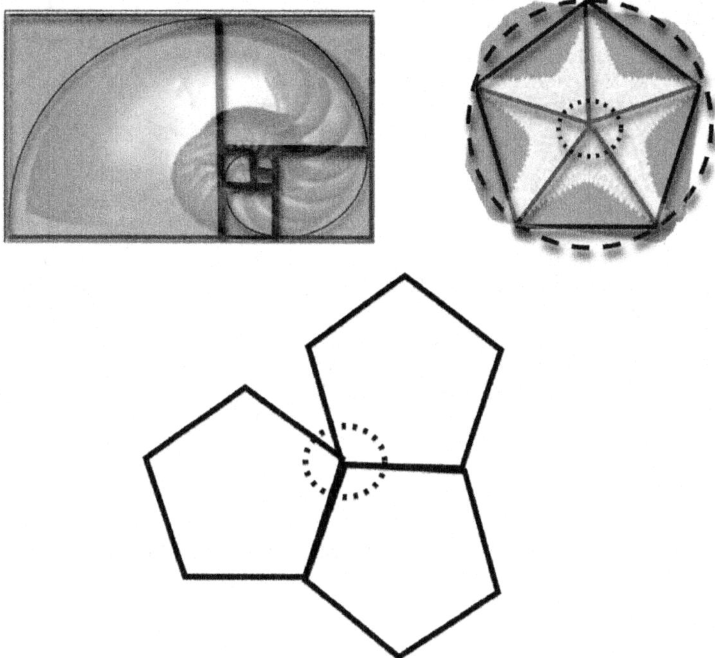

Figure 7.6   Pentagons don't tesselate, but they are a special shape.

SYMMETRY AS A MATTER OF CONGRUENCE

Congruence between two shapes means that everything about one shape is duplicated with same shape and measurements for the other shape. Symmetry within a shape means that everything on one side of that shape matches perfectly with the other side of that shape across an internal line of symmetry. Symmetry across an external line of symmetry means that the two images on either side of the line of symmetry *match exactly*. In Chapter 1, folding shapes was used as a strategy to identify lines of symmetry (the fold crease) within polygons.

The regular polygons in Figure 7.7 show folds as dotted lines on the polygons, so that one side of the shape folds to match exactly to the other side. Now it is time to use a circle to explore the same symmetry idea. Circles can be folded many times to match sides, and each fold crease also marks the diameter. Jean, a fifth-grader,[2] could not seem to remember all of the pieces and characteristics about circles (radius, diameters, circumference) from one day to the next. When she started folding a round coffee filter to help her get her hands and brain around symmetry and other circle elements, her understanding improved as well as her memory.

All of these creased folds representing internal lines of symmetry cross in the center of each polygon to generate central angles, a peculiarity of the regular shape. An outline of Jean's circle is also included in Figure 7.7 so that you can see what she saw with all of those folds and diameters and central angles of the circle. Many children, and adults too, are kinesthetic, hands-on learners who learn through their hands by

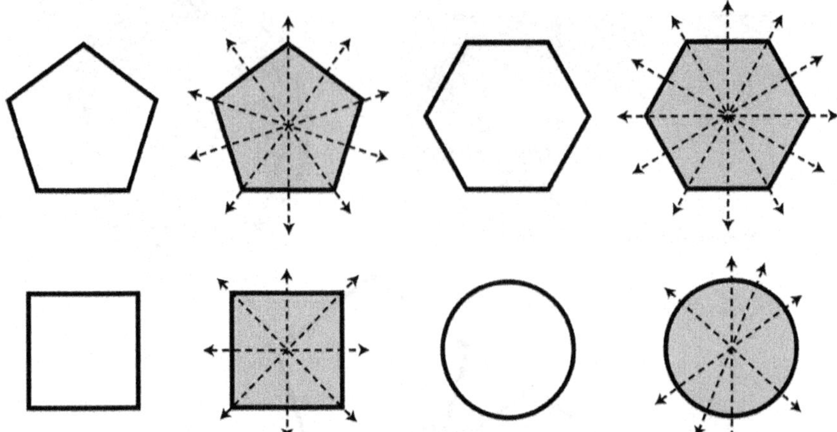

Figure 7.7   Fold lines of symmetry.

moving materials around to investigate relationships, cutting out pieces to see how they fit together, and, like Jean, folding shapes different ways to explore different ways of seeing things.

KEEP IN MIND

The examples in this chapter can appear in both elementary-level classrooms or in a more formal geometry class when a specific proof for congruence, similarity, or symmetry is required. However, when you are helping your child, your primary task is to encourage the visualizing and the strategic guessing. Your child will confirm their guesses at the appropriate grade levels—or you can help your child now with confirmations by using tracing paper and other hands-on math tools. Help awaken the Bombelli wild-guessing-aha-moments within your child's imagination!

NOTES

1. Hilton and Pedersen, *Fear No More*, 4.
2. Draper, *Winning the Math Homework Challenge*, 92.

Chapter 8

Special Spatial, Yet Irrational, Patterns

Mathematics is an activity governed by the same rules imposed upon the
symphonies of Beethoven, the paintings of Da Vinci, and the poetry of Homer.[1]

—Kasner and Newman

We memorize, and memorialize, pi as something of a rite of passage into middle
school or grade six, whichever comes first in Wherever-town, USA. Students are
asked to use pi to calculate a circle's area, sometimes without the slightest idea of
why it works so well. The memorized number is 3.14 (and sometimes the ratio 22/7 is
recruited to get through the course) and it "miraculously" allows calculations for areas
and circumferences of rounded shapes, most often circles. Neither 3.14 nor 22/7 are
accurate representations for pi (because both 3.14 or 22/7 are rational numbers; pi is
irrational), but they work for the short term to get through sixth grade.

This chapter mentions two special categories of numbers, irrational (cannot be
written as a ratio of integers) and transcendental (a very special number that is not an
algebraic number, but that definition isn't as important to you now as just knowing
that there is such a special category for numbers). The two categories of numbers are
mentioned here because we use pi (a member of both categories) frequently without
truly understanding how very special it is. Pi is awesome, as are the other rational and
irrational numbers, and it rarely gets the respect that it really deserves. Neither of these
categories will be on a chapter test, so sit back, relax, and try to enjoy.

This chapter explores that awesome yet irrational number, pi, that is used for calculating
areas and circumferences for those rounded shapes and solids, and some less familiar numbers, like phi (a number that fascinated the Greeks), "*e*" (a number used to explore limits),
and "$\mathcal{E}$" (a ratio that controls circles). The first three of these four numbers are both transcendental and irrational numbers. The fourth number is neither transcendental nor irrational, yet it plays an important role in mathematics. The special patterns in these numbers can
appear while comparing components of a shape or just comparing shape measurements.

Children need to use that special ratio pi (π) symbol so that they can calculate a
circle's area and circumference. They have probably been multiplying to find the areas
of certain polygons since the fourth grade, so some students, like Sally, may have

ventured into the nonpolygon arena of circles. Sally didn't use pi in her exploration (Chapter 4) but she did unknowingly start down that path into eventual limits. Knowing early on about how pi and phi are related to shapes, and why they work the way they do, helps children to understand how the other special numbers like *"e"* and *"Ɛ"* fit into the scheme of things when they are older.

PI, Π, AND 3.141592654 . . . —FROM ARCHIMEDES TO SUPERCOMPUTERS

Kim wrote the poem below about pi as part of a math project. She, like other linguistic learners, enjoyed writing poetry, especially when the poem could express a new depth of meaning, *and* provide a mnemonic for the digits in the process. The number of words per line in Kim's poem reflects the actual digits included in the pi ratio, and Kim even included the period used after the first line in the poem to serve as the decimal in the pi number. A poem that has this relationship of digits to words is called a "piem" and belongs to Piphilology study. What better way to start the journey into Pi-land—not with Homer's poetry but with Kim's piem.

I Never Change.
I
Start But Never End
And
Circle Is My Best Friend
For Greek Ratio I Am And I Am Incommensurable

Although mathematicians have continually worked toward finding a more accurate representation of pi, Archimedes is credited with having the best estimate in the second century BC. Three thousand years before Archimedes, early Egyptians used 3.1605 for their version of pi. Two thousand years before Archimedes, a Babylonian tablet showed 3.125 for the value of pi. Aryabhata (from Chapter 5 and triangle fame) continued the examination of pi with improved approximations three centuries *after* Archimedes' initial estimates. So, you see, over time, and now with the help of technology, mathematicians continue to close in on this famous number.

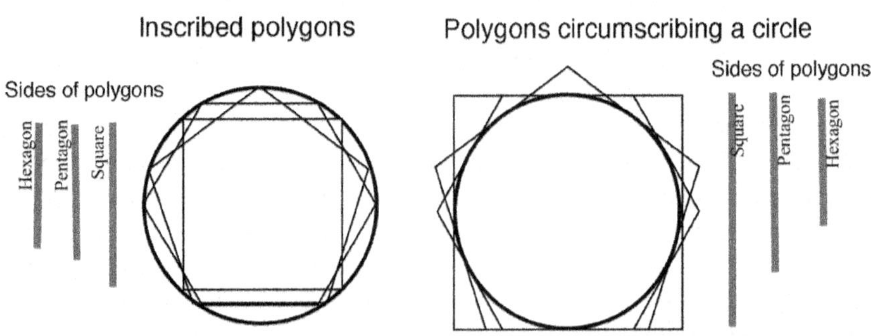

Figure 8.1   Archimedes' approximation for pi.

Today, many super computers are continually churning away so that the accepted pi ratio has now been calculated to 3.141592654 . . . continuing to ten trillion digits, however, the actual decimal representation continues without end. To get an idea of the magnitude of this number of digits, turn the 130 pages of David Slatner's 1997 book *Joy of Pi* to read a million of pi's digits *in very small type*—only 9,999,999,000+ left to get to the ten trillionth digit. Supercomputers calculate pi as an exercise, but most practical applications use only forty digits or so. Aren't you glad that your child may use the approximated 3.14 or 3.14159 to solve geometry problems?

Since we were not there when Archimedes worked on his calculations, all we have are history notes suggesting what Archimedes is said to have done. Figure 8.1 shows a visual diagram of how he used one circle to inscribe and circumscribe regular polygons, measured the perimeters of all these related polygon perimeters, and then compared each perimeter with the diagonal of that circle. Archimedes' goal was to get *very* close to a circle's circumference by closing in from the inside (inscribed) polygon and the outside (circumscribed—there is the "circum" prefix again!) polygon by increasing the number of sides of the regular polygons.

Archimedes used one circle and compared the polygon perimeters to the length of the circle's diameter. He noticed that as the *number of sides* of the inscribed polygon increased, the lengths of the sides would get shorter and the perimeters would increase, getting closer and closer to the circumference of the circle. The lengths of the sides have been pulled out from the inscribed polygons (and the circumscribed ones, too) in Figure 8.1 to make it easier to see how the lengths of the sides changed, getting smaller and smaller and closer and closer to the circle's circumference.

Add all of the sides of each inscribed polygon to get its perimeter and then divide that perimeter by the circle's diameter. Next, add the sides for each circumscribed *similar* polygon perimeter and divide by the same diameter. Archimedes is said to have repeated the calculation for the addition for perimeters and division with the circle's diameter with ninety-six pairs of polygons. Each division gets closer and closer to our now familiar number for pi as 3.14 . . . that still to this day is written as a continuing, nonpatterned irrational decimal with all ten trillion plus digits!

"Then [Archimedes] used the Pythagorean theorem to work out the perimeters of these inscribed and circumscribed polygons, starting with the hexagon and bootstrapping his way up to 12, 24, 48, and ultimately 96 sides. . . . This approach is known as the method of exhaustion . . . [by] exhausting the wiggle room for pi."[2] If you wish, you can replicate Archimedes work by adding and dividing until your polygons have ninety-six sides, or until you are exhausted, or at least satisfied enough to let the supercomputers take over.

For the tactile learner, use the ribbon or nonstretchable yarn from Chapter 4 and cut a length that is long enough to make a circle circumference. Cut another length for a diameter that measures though the center of your circle, as close as you can make it fit. Compare the lengths of the diameter and the circumference. Do this for several different sized circles and diameters. Look for a little more than three diameters to make the circumference. As the circles get larger, the diameters get longer, yet the ratio stays the same. Has your child ever wondered why the ratio didn't increase when both the circumference *and* the diameter increased? It's all about pi!

It turns out that a little more than three diameters of a circle are required to measure around the circumference (the sum of all of those sides in the polygons perimeters) of that same circle. Archimedes had approximated that ratio to between 223/71 and 22/7 or in decimal number form to between 3.14084507 and 3.142857143. Two millennia later an eighteenth-century British mathematician William Jones gave it the name "pi" and the symbol π that we use today. This approximated π ratio is used to calculate a circle area (the number of square units as described in Chapter 5) and circumference (number of diagonals linear units in Chapter 7).

PHI, Φ, AND 1.618 . . . —BEAUTY IN THE GOLDEN RATIO

Phi is yet another fascinating number that has piqued mathematicians' imaginations for centuries. It also happens to be another transcendental irrational number to whet your geometry pattern appetite. You may not have studied this ratio during your school days, but your child might see it today because more recent curricula standards are placing more emphasis on math patterns and relationships. Although phi is referred to as the Golden Ratio, it isn't a ratio that constitutes a rational number, but it (like pi) *is a comparison of two constructed measures* that generates a decimal without end. This phi ratio, like pi, has roots in early Greek culture.

Greek architects called phi the Golden Ratio, probably because of its pleasing appearance manifested in architectural design of the Parthenon and other Greek buildings. The length of the columns (long line segment) was in a ratio with a width between the columns (short line segment), generating a rectangle with one side 1.618 . . . times the adjacent side. The same comparison appears in the nautilus shell spiral (Figure 8.2) when a rectangle is drawn around the shell. The length of the long side of the rectangle is 1.618 . . . times longer than the shorter side, another reminder that nature is relationships in space and art creates relationships in space.

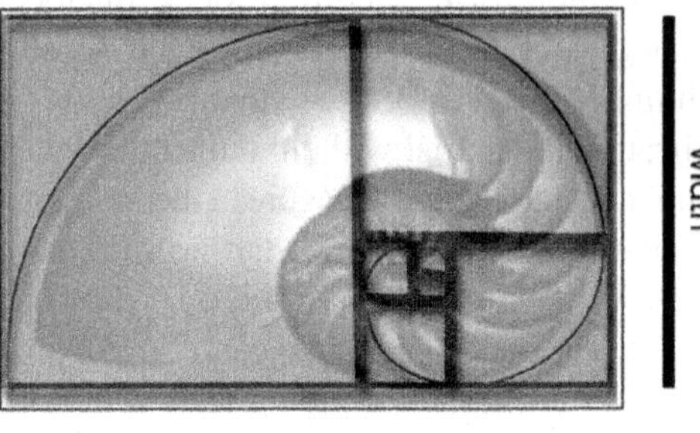

This length is 1.618... times the width

This length is 1.618... times the width

Figure 8.2   Phi ratio in a nautilus shell.

The Pythagoreans discovered phi in the pentagon-star relationship (see Figure 8.4), and later Johannes Kepler called it "one of the two treasures of geometry." In between the Pythagoreans and Kepler, thirteenth-century mathematician Leonardo Pisano

Bigollo, also known as Fibonacci, explored a wide range of mathematical relationships from prime numbers to pinecone patterns. The Fibonacci Sequence (1,1,2,3,5,8, 13, 21, ...) is a pattern that starts with 1 and continues forever (some sources have Fibonacci's sequence beginning with 0). The sequence self-generates each subsequent number by adding the previous two numbers.

Ask Yourself:

What relationship could Fibonacci's sequence have with the phi ratio?

Divide sequential pairs of numbers in the sequence using the second number in the pair as the numerator.   1, 1, 2, 3, 5, 8, 13, 21, 34, ...

$$\frac{1}{1}, \quad \frac{2}{1}, \quad \frac{3}{2}, \quad \frac{5}{3}, \quad \frac{8}{5}, \quad \frac{13}{8}, \quad \frac{21}{13}, \quad \frac{34}{21} \dots$$

1,   2, 1.5, 1.$\overline{66}$... , 1.6, 1.625, 1.6153...,1.619...

What kind of pattern do you see?

Figure 8.3   Task Box: Ask Yourself.

Current mathematicians and Math Aficionados delight, as did the Pythagoreans, in how a regular pentagon can regenerate a larger (and smaller) version of itself because of this ratio. The regeneration happens because of the recursive (repeating) operation: draw the regular pentagon and connect the diagonals so that a smaller regular

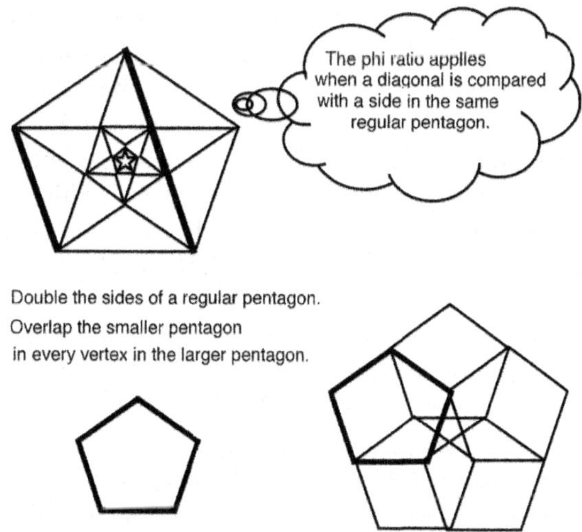

Figure 8.4   Pentagon phi and pentagon overlap. Do you see the same stars in both?

pentagon appears within the five-pointed star. Repeat drawing the regular pentagon and the diagonals like the one in Figure 8.4 to uncover the star. That is the recursive part. Yes, "recursive" means you do the same thing over and over again; it isn't insanity, just a pattern. The regular pentagon continues to alternate with a star formation.

Let's return briefly to the children's description of an image in Chapter 1, and examine the overlapping pentagon in Figure 8.4 through their eyes. The small pentagon is drawn in the larger double-sided pentagon and highlighted with a bold line. The other small pentagons in the large pentagon are not outlined, but you can draw them using a different color to highlight each small pentagon. The different colors can help you to see the tracked, and overlapped, pentagons to reveal a star with different color sides, each related to a small pentagon. Did you see the pentagons first or the star first? Do you see *both* now in the same diagram?

Just in case your child, or you, think that this Golden Ratio is dusty and occurs only in measurements in ancient Greek architecture or in the Pythagoreans' five-star pentagon design branding symbol for their fourth-century "club T-shirts," the twentieth-century architect György Doczi described some examples of how and where this ratio closely approximates measurements, from daisy petals to the human body and, yes, the side measures of that rectangular credit card in your pocket. In Doczi's words, this ratio is "a uniquely reciprocal relationship between two unequal parts of a whole."[3]

THE LIMIT OF *e* AND 2.71826 . . . —A CURIOUS NUMBER WITH CURIOUS COUSINS

Pi and phi are not the only comparisons that have a transcendental number status. Another transcendental number was given its label "*e*" by Leonhard Euler (pronounced "Oiler") in the sixteenth century, likely because of the first letter of his own last name. Possibly first noticed as a concept of compound interest but without the "*e*" label, the idea appeared on a seventeenth-century BC Mesopotamian tablet. About thirty-two to thirty-four centuries later, the idea resurfaced with other mathematicians—John Napier, (working with logarithms), Gottfried Leibniz (developing calculus using "*e*" as a limit notation), and Euler (describing mechanics).[4]

The number attributed to "*e*" came from that limit idea with an equation, $y=(1+(1/x))^x$. The brave, and maybe restless, can calculate this number, preferably with a calculator or even a spreadsheet, by replacing the "x" with positive values, and continue until you are exhausted, or are close to Euler's approximated 2.718266254 . . . number, whichever comes first. More descriptions about this limit idea appear in Chapter 12, but for now, use any calculator that has an exponent key, enter "(1+(1/20000))^20000" making sure to keep the parentheses in pairs. Euler was likely exhausted by the time 20,000 rolled around.

For those with a graphing calculator, enter "(1+(1/x))^x" in the first y = row after pressing the [y=] key, and then press the [graph] key to see the graph of the equation. Curious minds might want to know what the "y" values are for this line. Those number values are listed in the table for this equation (or you can press the [trace] key and put the curser on the line), so, by pressing the [2nd] key and the [tabl] key (just above the [graph] key), the values appear in the table for you to examine.

The limit for *"e"* is accepted as a number that is always 2.718266254 . . . (a decimal that keeps on going without a repeating pattern). The nonending and nonrepeating pattern for its decimal means that *"e"* qualifies for the irrational number club. It is indeed a curious number[5] and has "curious cousins" that incorporate *"e"* into their respective representations. One of those distant cousins is phi, provided that you trace the family-tree relationship through the family of logarithms, another reminder about the connectedness of concepts in mathematics. When that *"e"* replaces "10" as the logarithm base, it is renames the logarithm as a natural logarithm.

In Wells' *Curious and Interesting Numbers*, he lists six other transcendental numbers, several of them still in the "undecided category." One of these numbers is 1.234567891011112 . . . (note that the sequence in the decimal continuation is the sequence of counting numbers never to repeat) and another is e^π. Then there are those other numbers carrying people's names, such as Liouville's number and Mascheroni's constant, whose status as transcendental numbers are still in debate about their transcendental status. So you see, mathematics is certainly not a done deal and patterns are clearly alive and well, still argued and certainly explored!

ECCENTRICITY Ɛ—A CIRCLE DECIDER

You have seen π (pi), ϕ (phi), and *e* as transcendental and irrational numbers. A special ratio that is neither transcendental nor irrational, yet still significant, is called an eccentricity ($\mathcal{E}$) ratio. It serves as a "decider" for the conic section shapes and indicates how far away the conic section graph is from being a circle. Four of the conic sections (yes, they cut a cone like Paul's cuts in Figure 8.5) are named parabola, circle, ellipse, and hyperbola. The circle and ellipse stay contained finitely within the cone boundary so to speak, but the parabola and hyperbola cuts extend indefinitely along the edges of the cone.

These conic section slices of a cone appear in Figure 8.5 as a diagram along with sliced cones made in carpentry class by a high-school student named Paul. For those who find eccentricity controls interesting, the circle's eccentricity ratio is when $\mathcal{E} = 0$ and the ellipse eccentricity ratio has $\mathcal{E}$ values that are less than 1. The conic section parabola graph has a ratio of $\mathcal{E} = 1$, and the conic graph for a hyperbola has a ratio of $\mathcal{E} > 1$. These ratios are useful when describing the elliptical orbits of planets in our solar system.

Paul used a lathe to generate the cone shapes and sliced the cones at different slant angles so that the face of each slice represented one of the conic section graph curves. The faces of his sliced cuts are shown on the diagram sketch and identified as parabola (a curve with no end points), an ellipse (a curve that is more oval and bounded by the cone), a circle (a curve that is made by a horizontal slice of the cone), and a hyperbola (a pair of curves made by a vertical cut across both cones).

The eccentricity ratio compares the length of the respective major axis with the distance between the foci of the conic section graphs to a point on the circumference, a piece of information that may or not be interesting (or necessary) to your child or to

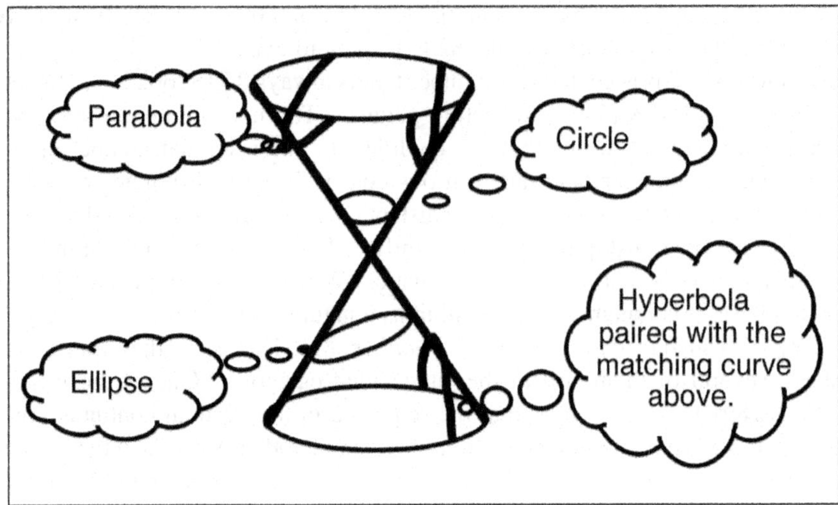

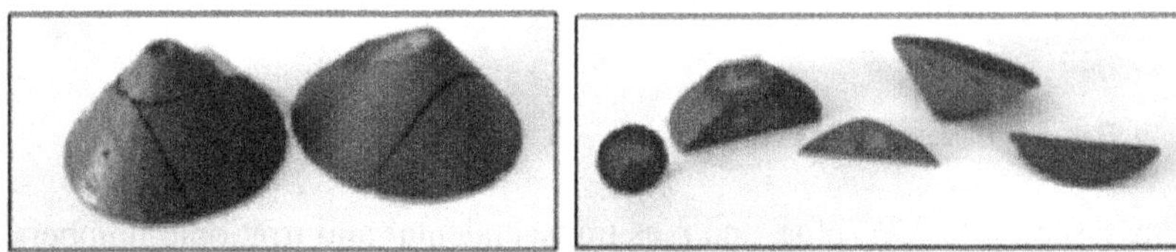

Figure 8.5 Paul's conic section cuts and cone diagram of slices.

you today, but it is *very interesting* to astronomers and astrophysicists. Math Aficionados enjoy studying shapes, numbers, and the relationships that each can tell about the other. That's where eccentricity comes in for your child; knowing about ratios that are not limited to just fractions in fifth or sixth grade will make comparing the conic sections distances in upper levels of math much more reasonable.

Science uses mathematics to examine natural events and patterns. Johannes Kepler, an astronomer, was specifically interested in the elliptical shape of the plant orbits based on earlier work by Copernicus. With elliptical shapes, the eccentricity ratio measures how far away the roundness of an ellipse is from being circular. Different ellipses have different eccentricity ratios and different shapes. To clarify a question in Chapter 7, all ellipses are not similar, just as all triangles are not similar. A good question could be, "do all of the planets in our solar system have the same eccentricity orbit?" What would happen if they did?

Ask Yourself:

Which eccentricity ratio for the orbits of the planets
is closest to being circular?

Earth 0.017	Mars 0.093
Venus 0.007	Saturn 0.056
Jupiter 0.048	Neptune 0.009
Mercury 0.206	Uranus 0.047

Which is farthest away from a circle? Ask some of
your own questions.

Remember, it is okay to ask a question
without knowing the answer.

Answer:
Venus

Figure 8.6   Task Box: Ask Yourself.

KEEP IN MIND

Numbers are "curiouser and curiouser" when compared in special ways and put under the microscope of comparisons and examinations. Who would have imagined that a simple circle, or regular pentagon or nautilus shell, could hatch such fascinating components and transcendental status? How many transcendental numbers are there? The answer is still a work in progress, although Georg Cantor is credited with proving that almost all numbers are transcendental, and that there are as many transcendental numbers as there are Real numbers (but who's counting?). Some numbers are "curiouser" than others, but then, fascination is in the eye of the beholder.

NOTES

1. Kasner and Newman, *Mathematics and the Imagination*, 362.
2. Strogatz, *Joy of X*, 127.
3. Doczi, *The Power of Limits*, 13.
4. Maor, *e: The Story of a Number*, 23.
5. Wells, *Curious and Interesting Numbers*, 37.

Chapter 9

Shape Changes and Patterns

Once one understands the pattern with some certainty, then it becomes clear what to do about it—and what not to do about it.[1]

—Thomas West quoting Benoit Mandelbrot

Mathematics with certainty? What a concept! Math Avoiders would like to have that kind of certainty to know that mathematics at its best is predictable. Because of this predictability, math patterns can provide some certainty about what to do and what not to do for you too. Your child can have increased clarity about what to do, and what not to do, by asking questions. This chapter is about developing that clarity as it applies to shapes. Changes in shapes will affect their respective perimeters and areas, but what happens with shapes when you keep the perimeter or area constant? What happens to areas when you start changing side lengths?

These are the kinds of questions that Math Aficionados like to ask and then sketch shapes to try to guess some possible solutions through patterns in the search for clarity. Asking questions can help you and your child focus on specific relationships. Use your best Pòlya or Halmos or Bombelli-type guesses. Sometimes there is one answer and sometimes there are several answers, and all of them can be accurate. It all depends on the specifics in the question. Asking questions is how you can stand on the shoulders of giants. Before you know it, you and your child will be able to sketch and guess and answer some of these questions yourself.

All you need is a pencil, some blank paper, curiosity, imagination, and a healthy appetite for asking some searching questions in order to gain more clarity. Good questions will connect different topics. You have probably heard the phrase "there is no such thing as a dumb question," and the phrase likely made you breathe a sigh of relief. Having the courage to ask these questions is proof that you are thinking mathematically, just as did famous mathematicians Pòlya or Halmos or even Bombelli of "wild thoughts" about imaginary numbers fame. By the way, "I don't get it" may be a declaration of frustration or a plea for help, but it is not a question!

SAME PERIMETERS, SAME AREAS, DIFFERENT POLYGONS

Start with two very different shapes. Is it possible for a pentagon to have the same perimeter as a triangle? You may want to try a few sketches yourself before reading on. Is it possible for two different rectangles to have the same perimeter? For example, can two different rectangles both have 12 as a perimeter? Do the sides always have to be whole numbers? Figure 9.1 shows several examples of different shapes with the same perimeter. If different shapes can share the same perimeter, then is it possible for two different shapes to have the same area? For example, a rectangle can have an area of 20 units, but can a square also have an area of 20 units?

   Your child can learn about perimeters by using different rectangles, coloring them, and then comparing them like Katrina did in Chapter 5 or by cutting them out like Sasha did, after coloring the edges for perimeter and the interior of the rectangles for the areas (Figure 9.2). Sasha was exploring all of the different rectangles that had

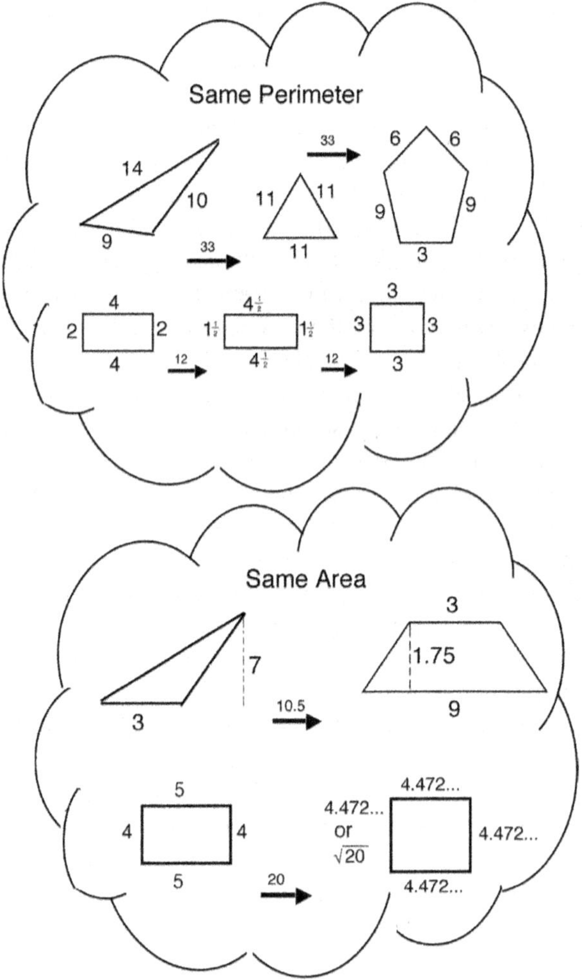

Figure 9.1   Same perimeter; same area.

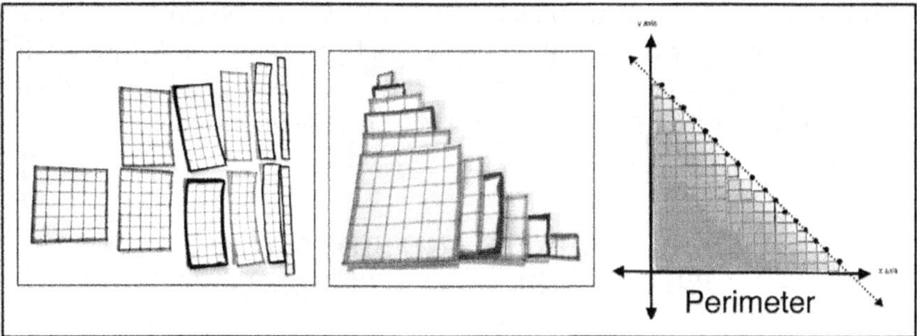

Figure 9.2 Sasha's perimeter experiment.

whole number side measures with 24 as a perimeter. Sasha was old enough to compare the shapes with the same perimeter, whereas Katrina was only trying to distinguish between the perimeter and area definitions.

Sasha first ordered the rectangles to see how the lengths got longer as the widths got shorter. Then she stacked them to show another relationship, making sure that one corner of all of the rectangles shared the same location. Do you think that your child will see what Sasha saw in Figure 9.2? There are several ways to describe these relationships. Young children can make a bar graph with the rectangles in the first photo, older children can stack the rectangles to see another relationship, and algebra students might recognize the graph of the arranged perimeters as 2x+2y=24 or y=-x+12.

Sasha used only positive integers for side measures, but if your child asks about other numbers, then by all means allow them to use fractions (or square roots or decimal forms of numbers). The relationship among the area of rectangles presented a different relationship. Sasha, and your child can as well, cut out all of the rectangles with an area of 36, and then she used different colors for the rectangles. She stacked the rectangles like the ones in Figure 9.3 being careful to align the lower left corners. Again, the rectangles can be arranged to make a bar graph or represent the graph of the stacked areas as x•y=36.

Do the dimensions of these rectangles remind your child of factors? Of dividing? Of square roots? Of prime numbers? Does your child recognize factors by looking at lengths and widths of areas? What about the pattern with the perimeters? The areas? Would you or your child have imagined that the perimeters (Figure 9.2) would show

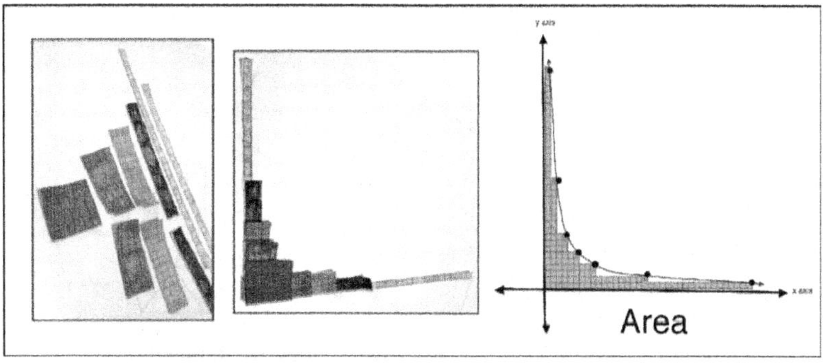

Figure 9.3 Sasha's area experiment.

a different curve than the areas (Figure 9.3)? How would the curves compare if the same perimeter number is used for both area and perimeter (24 for area and 24 for perimeter)? So many questions are just waiting to be asked and investigated when children actually get their hands on the mathematical relationships.

SIMILAR SHAPES AND SIDE CHANGES PREDICT AREAS AND PERIMETERS

When your child compared similar shapes, they were required to keep faithful to a proportion to make sure that the similar relationship didn't change. Now you can use that idea and take it another step to compare what this proportional change can do to areas and perimeters. For example, if all the sides of a shape are doubled or tripled, then what does that enlargement do to the relationship of the areas of these similar triangles? What would that do to the perimeter? The similar shape relationship is maintained since "doubling" is a 2:1 proportion, but does the area also have the same 2:1 proportion? How about the perimeter? Read on.

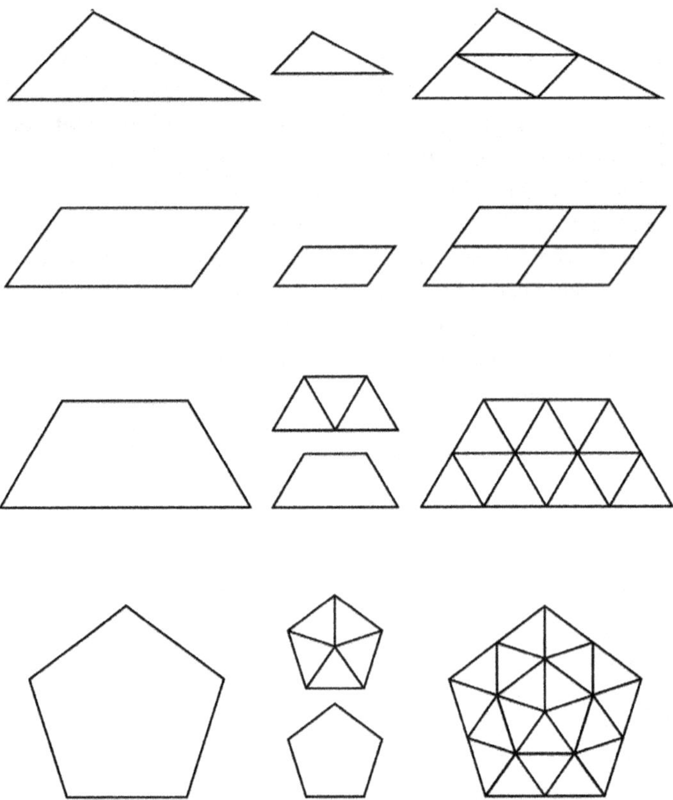

Figure 9.4   Decomposing shapes to compare areas when sides are doubled.

Flexibility with composition and decomposition comes in handy as your child draws pairs of similar shapes, like the ones in Figure 9.4. Each paired shape represents the "before" and "after" of the sides being doubled, along with what the small shape looks like within the larger shape. The trapezoid and the pentagon and other polygons do not overlap within the larger double-sided similar shape[2] as nicely as the triangle and the parallelogram, so those shapes may need to be decomposed into triangles in order to recognize the relationship between doubling the sides and the resulting areas.

This side-to-area relationship is a reminder of how a Linelander's side has one kind of measure (one-dimensional line segment) and a Flatlander's area shape has a different kind of measure (two-dimensional square units). The line segment Lineland sides (one dimension) are simply doubled in length but the shape area (two dimension) in Flatland has a different unit. Doubling (x2) a side means that the perimeter would double (count the sides to confirm this), but the area would be quadrupled (x4 or 2^2). What do you suppose the relationship between sides and areas would be if you tripled the sides of these same shapes? Start drawing.

How would changing the sides lengths affect the volumes (and faces) of three-dimensional polyhedra? Would the same relationship of 2:1 proportion of side increases carry over from area to be a 4:1 proportion for volume? These kinds of questions almost demand a blank sheet of paper with a pencil in order to sketch some spatial renditions of the polyhedra. Having the imagination to assign some numbers to the sides (these are called edges in polyhedra) will help jumpstart some data collection to improve clarity. Curiosity and persistence will get you through the what-to-do and what-not-to-do phase of the investigation.

DOTS, VERTICES, SPACES, AND FACES

One particular shape, the pentagon, has intrigued mathematicians' imaginations since the Pythagorean's image brand. In Chapter 8, the lengths of the lines of the star inside the pentagon were compared in a ratio that gave us phi, 1.618 . . . , a number that continues to reappear in a wide variety of settings, possibly most notably in da Vinci paintings. The pentagon (and other shapes) gives up another interesting number relationship among the number of line intersections, the number of spaces, and the number of line segments.

"'Now let's make a few dots in our star,' continued the number devil. 'Place one at every point where the lines cross or come together. Count how many there are.'"[3] This is from a story about Robert's dreams described in a math fantasy book titled *The Number Devil*, and no, they are not nightmares. The number devil is regularly appearing in Robert's dreams to show how often mathematics can show up in different and interesting places with a variety of curious relationships. Robert is dreaming and Alice fell down the hole chasing a rabbit. You never know where you are going to find some mathematics!

In his tenth dream, the number devil presented a star inside the regular pentagon (we explored another relationship between the star and the pentagon in Chapter 8)

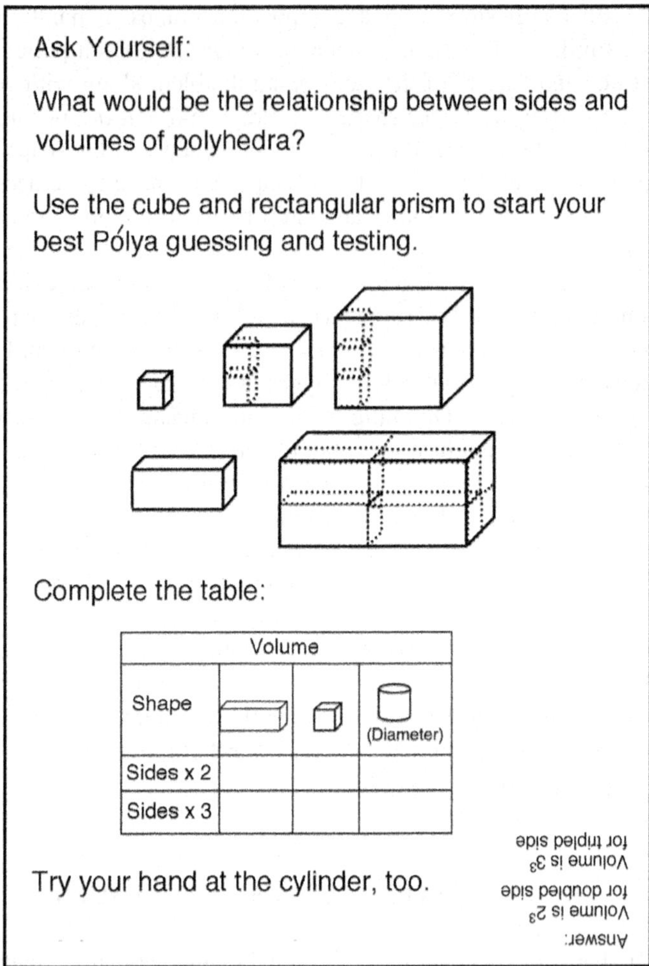

Volume			
Shape			(Diameter)
Sides x 2			
Sides x 3			

Figure 9.5   Task Box: Ask Yourself.

to Robert so that Robert could follow the number devil's directions. Together, they investigated a relationship among the number of dots, spaces, and line segments in different shapes shown in Figure 9.6 (this time convex is not required). As they drew new arrangements and counted each shape's dots, spaces, and line segments (even though they said lines in the equation, they knew that they were really *line segments*), a pattern started to emerge. Does your count match Robert's count?

They found an interesting number relationship between the dots, spaces, and line segments, so the number devil reminded Robert that the relationship worked for three-dimensional solids, too, just with a dimensional twist. Leonhard Euler (pronounced like "Oiler") and the number devil may not have crossed paths (pun intended), but they do share a common interest in patterns. Euler figured out that a relationship existed among the faces, edges, and vertices of any polyhedra, very similar to the number devil's dots, spaces, and lines in Robert's dream. The dots from the two-dimensional

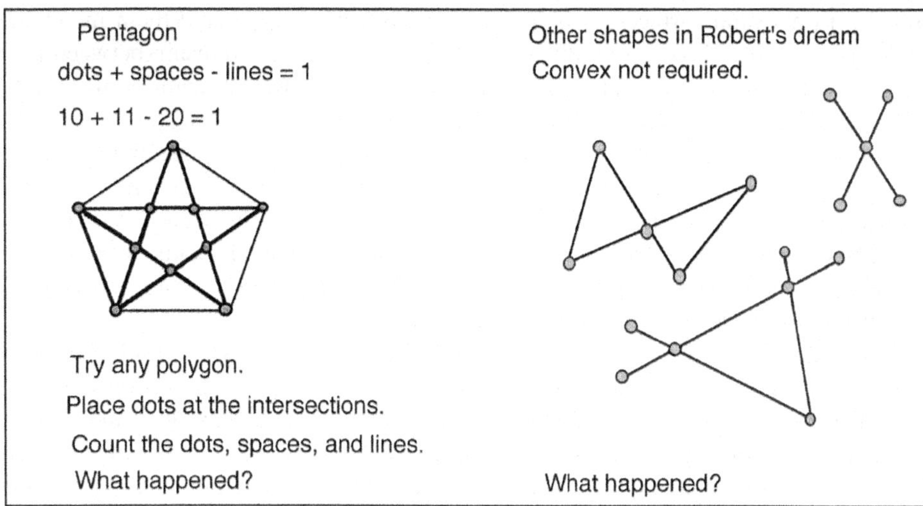

Figure 9.6   Number devil's work with a pentagon and other arrangements.

shapes become the vertices in the three-dimensional solids, the line segments become the edges, and the spaces are replaced with polyhedra faces.

Your child might have seen Platonic solids (Figure 9.7) in the fifth grade and learned that these polyhedra solids *must have specific relationships* to qualify for the Platonic Solid title. You read about them in Chapter 6 as three-dimensional solids with

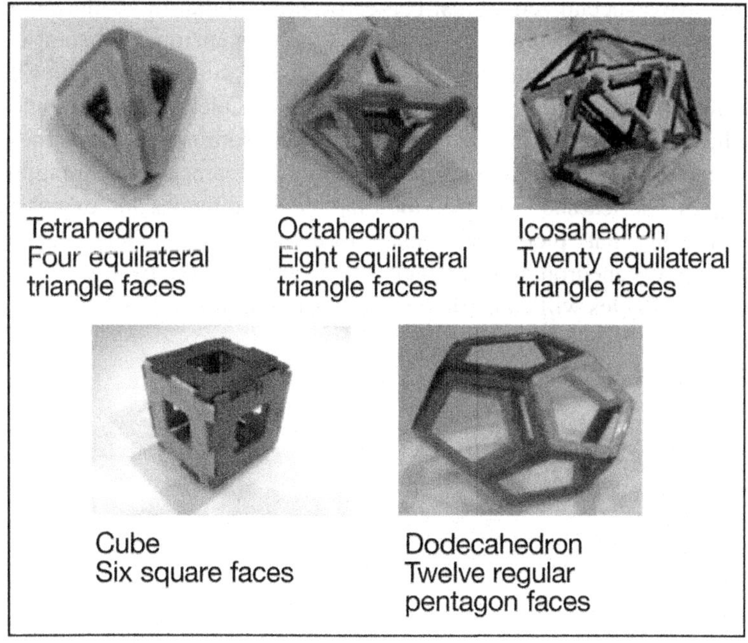

Figure 9.7   Platonic solids.

polyhedra nets and surface areas. You might have even discovered them in science class as minerals. Now, here they are again with a pattern relationship between their faces, edges, and vertices that Euler ("Oiler") found over twenty centuries after Plato, another reminder that mathematics is not a speed sport!

There are only five Platonic Solids because the regular same-shape and congruent faces that are the polyhedra's net *must* fold up to make a solid shape *and* have congruent angles at all vertices *and* have the same number of faces at each vertex. Plato must have been so awed by these polyhedra that he attributed elements of the cosmos to specific solids: cube for earth, tetrahedron for fire, octahedron for air, icosahedron for water, and dodecahedron for the Universe itself.

While the fate of the cosmos may not be at stake for you or your child, the notion that there are *only* five solids that fit these criteria is indeed interesting! Euler (a reminder to say "Oiler") carried the polyhedra pattern idea a bit further and found a consistent relationship that existed among the faces, edges, and vertices of *any* polyhedra, not just the Platonic solids. Euler may not have been the only one to recognize this relationship, but he did claim naming rights. His name is attached to the relationship, F+V-E=2 (can also be V+F-E=2. Do you know why?). The selection of solids in Figure 9.8 satisfies his (and the number devil's) equation.

This equation relationship translated into words states that "the number of faces (F) in any polyhedra" can be added to "the number of vertices (V) in the same polyhedra"; then, following the number devil pattern, subtract the number of edges (E) "the number of edges E of the same polyhedra" and that total will always equal 2. Euler established that this relationship applied to all solids, past and present, and would also apply to any future twentieth-century solids, including Norman Johnson's ninety-two specialized solids. Have you noticed that the pattern equation for the two-dimensional constant is "1" and the three-dimensional constant is "2?"

Teresa, a college student, took Plato's Platonic Solids and Euler's (another reminder that his name is pronounced to sound like "Oiler") pattern to another level by using her poetry to describe the relationship. The math project guidelines specified that the students would incorporate their learning preferences and strengths with the math that they had learned. She used the Platonic Solid names and expressed Euler's relationship using that "piem" relationship of numbers. Her radiolaria skeletons title represents the icosahedron and the twelve lines added to the twenty words and subtracted thirty syllables will equal the two words in the title.

Radiolaria Skeletons
Under
The sea
Microscopic
Animals
Hiding from you
Hiding from me
Shaped from
The triangles
Caught
On the
Ocean's
Floor.

Platonic Solids	Faces	Vertices	Edges	Euler's relationship
Tetrahedron	4	4	6	$4 + 4 - 6 = 2$
Cube	6	8	12	$6 + 8 - 12 = 2$
Octahedron	8	6	12	$8 + 6 - 12 = 2$
Dodecahedron	12	20	30	$12 + 20 - 30 = 2$
Icosahedron	20	12	30	$20 + 12 - 30 = 2$

Six of the 13 Archimedean Solids	Faces	Vertices	Edges	Euler's relationship
Truncated tetrahedron	8	12	18	$8 + 12 - 18 = 2$
Truncated cube	14	24	36	$14 + 24 - 36 = 2$
Truncated icosahedron	32	60	90	$32 + 60 - 90 = 2$
Cuboctahedron	14	12	24	$14 + 12 - 24 = 2$
Icosidodecahedron	32	30	60	$32 + 30 - 60 = 2$
Rhombicosidodecahedron	62	60	120	$62 + 60 - 120 = 2$

Seven of the 92 Johnson Solids	Faces	Vertices	Edges	Euler's relationship
Square based pyramid	5	5	8	$5 + 5 - 8 = 2$
Pentagonal pyramid	6	6	10	$6 + 6 - 10 = 2$
Pentagonal cupola	12	15	25	$12 + 15 - 25 = 2$
Elongated square gyrobicupola	26	24	48	$26 + 24 - 48 = 2$
Triangular bipyramid	6	5	9	$6 + 5 - 9 = 2$
Gyrobifastigium	8	8	14	$8 + 8 - 14 = 2$
Augmented triangular prism	14	7	21	$14 + 7 - 21 = 2$

Figure 9.8 These solids share Euler's relationship.

KEEP IN MIND

Although there are many more patterns and relationships to explore in the geometry of space, you and your child have completed quite a few if you have read this far in this book. Thomas Edison said it best: "Eyes that look are common. Eyes that see are amazing." Let your eyes see some amazing things! He is also credited with saying that he knew ten thousand ways to *not* build a light bulb. Do not let your amazing eyes be dimmed by things that don't work the way you want the first time you try an amazing idea on for size.

NOTES

1. West, *In The Mind's Eye*, 289.
2. Draper, *Winning the Math Homework Challenge*, 85.
3. Enzensberger, *The Number Devil*, 202.

Part IV

CONNECTIONS

After all, mathematics is everywhere around us if we know how to recognize it.[1]

—Theoni Pappas

Do you believe that mathematics is everywhere? Does your child? To Math Aficionados, mathematics *is* everywhere and they *can* see it. Like Pappas, they can see all those relationships and connections in and among the shapes! To the Math Avoider who probably does see the shapes but not the connections and interrelationships, neither is it recognizable nor is it believable as evidenced by the wailing cry, "When will I ever use this?" Joe, a seventh-grade student, expressed his idea of math applications and logic in his math journal.

> The thing that most people don't realize about math is that all math really is logic. Pure, unadulterated. Valuing the application of math can only come if you can apply logic—and I don't think you can give logic a value. Logic is applied to reality in our everyday lives, and all math does is gives [sic] it universal symbols, numbers, and characters.

When you attempt to answer your child's wailing question with justification examples from applications, you are addressing only the symptoms of the issue, not the cause. Just stating that it is about logic doesn't help either, even though Joe believed strongly in the logical system behind the math. The cause for the complaints and wails is confusion. You can help resolve your child's confusion by showing *how* topics fit together and *why* they fit together.

The central focus of this section is to show in Chapter 10 how the right triangle strongly influenced the Pythagorean Theorem and trigonometry. The focus of Chapter 11 describes the connections that the circle has on many geometry relationships. Chapter 12 brings in several connections to weave in and through how your child's early learning continues to reappear and even shows up in the pedestal-sitting math topic called calculus.

By now, if either you or your child were in that wailing chorus when you first started this book, then the cry isn't as loud as it used to be or maybe it has even reduced to almost a whisper. Maybe it has even disappeared? To paraphrase Auntie Mame's comment on life from the 1958 film bearing Auntie Mame's name, "There is a banquet of connections out there ready to explore in mathematics and too many Math Avoiders are starving." You and your child can enjoy the mathematical connections banquet!

The National NCTM Math Standards describes three categories for identifying mathematical connections: within mathematics, between mathematics and daily life, and between mathematics and other disciplines. To add to the other previous math-connection examples in this book, all three kinds of math connections are addressed in this section: within mathematical topics, across a variety of everyday kind of situations, and mathematics as it applies to other disciplines. The Pythagorean Theorem and trig connections provide some jaunts into everyday examples.

NOTE

1. Smith, *Agnesi to Zeno*, 195.

Chapter 10

The Influential Right Triangle

When Thales visited Egypt and used the method of shadow-reckoning to find the height of the Great Pyramid, . . . we are told that he astonished the Egyptians.[1]

—Kasner and Newman

Imagine building a 455-foot tall pyramid without the benefit of heavy cranes or other modern skyscraper-building machinery. Any tool that those early builders used needed to be accurate enough so that all four sides of this pyramid reach the same apex or the whole thing was a mistake. Those builders are said to have tied evenly distributed knots on a rope to make a right triangle so that they would know where to place the stones from the ground up. The Egyptian "rope stretcher surveyors" may have measured a 90° angle by using a geometric cross-staff similar to our current day T-square. Either way, that right angle made the sides meet at the apex.

This chapter's topics describe how influential that right triangle is, as it gradually progresses from just being one of the triangle shapes in the first grade to compositions of shapes beginning in the third grade. Eventually in the middle grades, the right triangle topics bloom into the Pythagorean theorem, when all of those descriptions and properties come together into Johannes Kepler's description as one of geometry's great treasures (the other one was phi) that "we may compare with a measure of gold."[2] The treasure doesn't end in middle school; it continues into trigonometry and calculus, and into Buzz Lightyear's beyond!

SHADOWS, RIGHT TRIANGLES, AND PROPORTIONS

Thales astonished the Egyptians because he could calculate the height of the right triangle that he "saw" in the pyramid height. Whether the height had been forgotten over the few thousand-year interval or had never been known, Thales resolved the issue by placing a stick in the sand and measuring the shadows of the stick and the pyramid *at the same time of day*. Thales could "see" the right triangles made by the pyramid's shadow and the stick in the sand and he shadow-reckoned that the two triangles had

99

the same angle measures. The mathematics of proportions and measurement allowed him to figure out the pyramid's height based on the stick.

Thales recognized that the pyramid's height was part of a right triangle (Figure 10.1) with one leg as the unknown height of the pyramid and the other leg was the shadow length plus half of one side of the pyramid's square base. Because the pyramid's right triangle was similar to the stick's right triangle, he could use the measures of the legs in the small stick right triangle made by the stick and its shadow in a proportion with the shadow leg of the pyramid to calculate the unknown height. He used what he *could measure* to calculate what he *couldn't measure*.

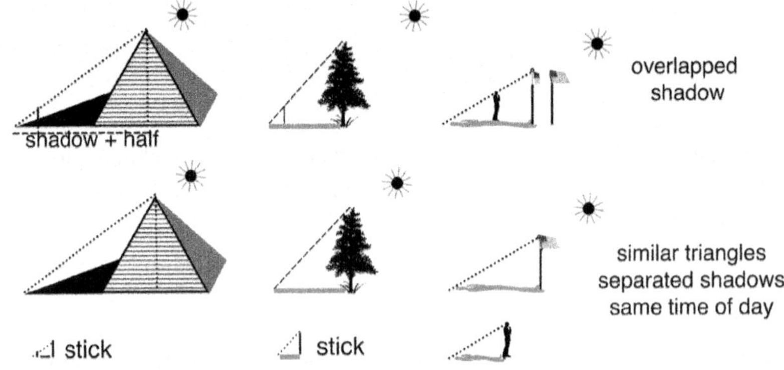

Figure 10.1   Right triangles and shadows.

If a flagpole, tree, or Thales' pyramid is too tall to actually measure, then your child can use the lengths that they *can measure* (Thales' stick in the sand, a person's height, and shadow lengths) to calculate the heights of tall things without the scratches and bruises of climbing any of them. People, poles, trees, and pyramids stand perpendicular to the ground (at least locally speaking, thanks to gravity) so the right triangles made by those overlapping shadows of people, trees, poles, and pyramids *at the same time of day* must be similar triangles. If the triangles are similar like the ones in Chapter 7, then the sides are proportional.

A quick reminder from Lizzie's and Theo's experiences described in previous chapters: the sides of triangles can shrink and grow, but the angles must stay the same for the triangles (or any similar polygon or polyhedra) to be similar *even when they are in a different location*. Judging by the sun casting shadows in Figure 10.1, the shadows for the pole and the person (the stick and the tree or pyramid) must be measured at the same angle (at the same time of day) to keep the triangles in a similarity and therefore proportional relationship.

THE PYTHAGOREAN THEOREM: A RELATIONSHIP OF SIDES

Ever heard the question, "What do President James Garfield, Leonardo da Vinci, and Bhaskara have in common?" Answer: all presented a geometric proof for the Pythagorean theorem. In fact, they aren't the only ones. An unknown Chinese author, or maybe several authors, described the earliest known *text explanation* (circa 600 BC Chinese) of a Pythagorean triple (right triangle with integer sides, such as 3, 4, 5). An earlier use appears in the 2000 BC astrological arrangements of Stonehenge in England.[3] These examples predate the Egyptian pyramids by several thousand years, and Pythagoras by centuries. Talk about staying power!

Pythagoras himself may have learned about it when he traveled to Babylonia and brought it back to Samos *before* it was designated as the Pythagorean theorem. Identifying the theorem with Pythagoras' name was possibly generated as an honor bestowed on him by his dedicated Pythagorean followers after Pythagoras actually proved the theorem. Euclid included a formal definition of this theorem in his *Elements* several centuries after Pythagoras, thus cementing its importance as part of the overall geometry structure. During the intervening years, other mathematicians and interested Math Aficionados played around with more proofs of this theorem.

In the seventeenth century, Pierre de Fermat hypothesized that the expression would *only* work for the positive integer "2." Being a mathematician who wanted to be sure he had left nothing to question, his equation was $x^n+b^n=c^n$ so that "n" represented an integer, a notation that just about covered all possibilities. The trouble was that Fermat only alluded to a proof in the margins of his book so other mathematicians couldn't examine or confirm it. It took another three centuries for Andrew Wiles, a British mathematician, to prove that indeed Fermat was right all along. One more example of mathematicians "standing on the shoulders" of previous giants!

Elisha Scott Loomis joined the Pythagoras, Euclid, Bhaskara, da Vinci, Fermat, Byrne, Garfield, and Wiles crowd with two volumes presenting 370 different proofs for the Pythagorean theorem, each with a different picture to show how the area arrangements worked. By the way, the Pythagorean-theorem relationship is guaranteed to work with both inferences: leg relationship guarantees a right triangle and a right triangle guarantees the leg and hypotenuse relationship.

Said another way, if the two legs, "a" and "b," of a right triangle have the Pythagorean relationship with the hypotenuse "c" so that $a^2+b^2=c^2$, then a right triangle is guaranteed; also, having a right triangle guarantees that the legs of that triangle have the $a^2+b^2=c^2$ relationship with the hypotenuse. Having such a guarantee with both conditions, right triangle first or the relationship first, is very special in mathematics.

You might remember the theorem formula as it applies to a right triangle or you may remember that it was just another formula that you had to memorize in order to pass the test or get through the coursework. Be assured that your child will also have to learn to use the Pythagorean theorem, so let your child know that remembering and applying the relationship is really about the relationship among the *areas of similar shapes that are generated from those special side measures*. The Pythagorean theorem trig ratios (see later in this chapter) can also be used to calculate the heights of the tree, pyramid, and flagpole, given the appropriate measures.

There is something rhythmical about the recitation of "*a* squared plus *b* squared equals *c* squared" that your child probably easily remembers. However, the phrase has as much to do with *areas of similar shapes* as they have with the side measures. This misunderstanding becomes apparent when your child says, or reasons, that if $a^2+b^2=c^2$ in a right triangle, then a+b=c must also be a true. If your child made the triangles with the straws in Chapter 3, they have already disproved this. Two sides of any triangle can *never* equal the third side, *ever*. If they do, then the triangle will no longer exist because the internal space of the triangle collapses.

Generally, the diagrams of the Pythagorean theorem relationship show areas with squares, illustrated like the ones in Figure 10.2. Areas of similar shapes *that are not squares* (but have areas calculated in square units) can also share the Pythagorean relationship for side measures, as long as they conform to the side measures that make a right triangle, also shown in Figure 10.2. Remember, the square unit is a measuring unit for all shapes in Flatland. Not just any willy-nilly areas will work; the areas depend on the right triangle side measures. Calculating those areas by decomposing shapes are explained in Chapter 5. And formulas work too.

Let's go back to the rope example possibly used for the kings' pyramids in Egypt. If the pyramid builders tied evenly spaced knots on a long rope, including the two ends, the spaces between the ropes would likely have been three spaces, four spaces, and five spaces to form a right triangle (when folded to connect the end knots as one) providing the desired right angle. Because the Pythagorean theorem relationship works so nicely with whole numbers, the 3, 4, and 5 being one example, the Pythagoreans called them Pythagorean triples, and in the process, figured out other numbers for triples (like the ones in Figure 10.3) to make right triangles.

Put on your best algebra-substitution gloves, and replace the x, y, and z labels of that last triangle in Figure 10.3 with the numbers from the table. These substitutions convert that triangle into yet another Pythagorean triple triangle. Just be sure to use the numbers in their relative length positions on the triangle: the longest measure (greatest number) is assigned to the hypotenuse where the "x" is, the shortest measure (least number) replaces the "z," and the middle measure (number in-between) is used instead of the "y" on the triangle.

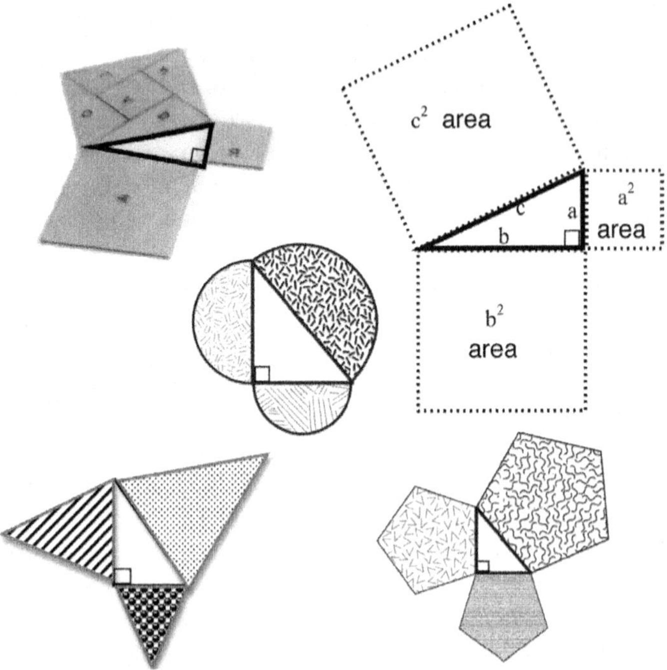

Figure 10.2 Pythagorean theorem is about areas of other shapes, too.

3, 4, 5	6, 8, 10	9, 12, 15
5, 12, 13	10, 24, 26	15, 36, 39
7, 24, 25	14, 48, 50	21, 72, 75
20, 21, 29	16, 30, 34	24, 45, 51

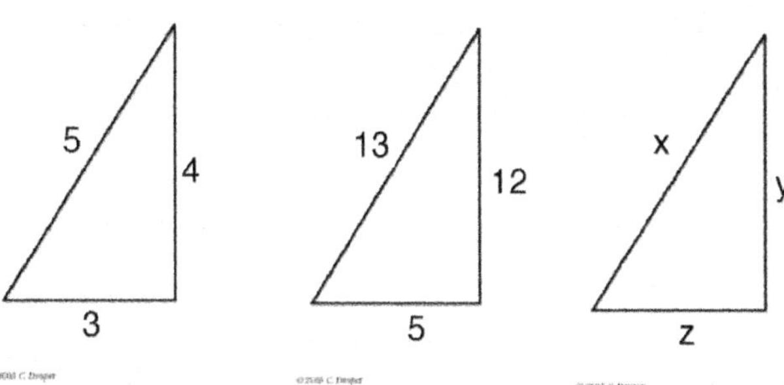

Figure 10.3   Pythagorean triples.

More recent Math Aficionados and mathematicians have stood on the Pythagoreans' shoulders and developed some formulas that would generate Pythagorean triples. Remember, no fractions are allowed for triples. The trick is to select the *positive integers* that conform to the restrictions (and limitations) included with these formulas. For example, choose a small number for "k" (to keep it simple) for the first formula in the task box in Figure 10.4, and then choose the other two numbers for "n" and "d" so that the number you choose for "n" is greater than the number that you choose for "d." Pull out that calculator so that most of your time is spent on thinking.

Try This:

Generate your own Pythagorean triples with these two formulas:

1. For n > d, select numbers for k, n, and d. Use 2knd, $k(n^2 - d^2)$, $k(n^2 + d^2)$

2. For n > 3, n is odd, and n is the shorter leg. Use $\dfrac{n^2 + 1}{2}$ for the hypotenuse and $\dfrac{n^2 - 1}{2}$ for the longer leg.

Be sure to use only positive integers.

Sample answer:
Choose k=1, n=5, d=2. Substitute
You will end up with 20, 21, and 29 so that
$20^2 = 400$, $21^2 = 441$, and $29^2 = 841$.

Figure 10.4   Task Box: Try This.

Regina, a middle-school student, found another way to examine how much creative imagination the early Math Aficionados must have had to identify this relationship. She did not use the squared number for a^2 as a square shape in her exploration. She reasoned that just because the a^2 *could* be arranged into a square, it did not *always have* to be a square. First she arranged the squared numbers, 3x3 as nine and 4x4 as sixteen, in rows like the circles in Figure 10.5. Then she made another row of circles to correspond to 5x5 for the twenty-five circles. The nine circles and the sixteen circles matched to the twenty-five circles. Regina's picture diagram cleared up her wordy explanation.

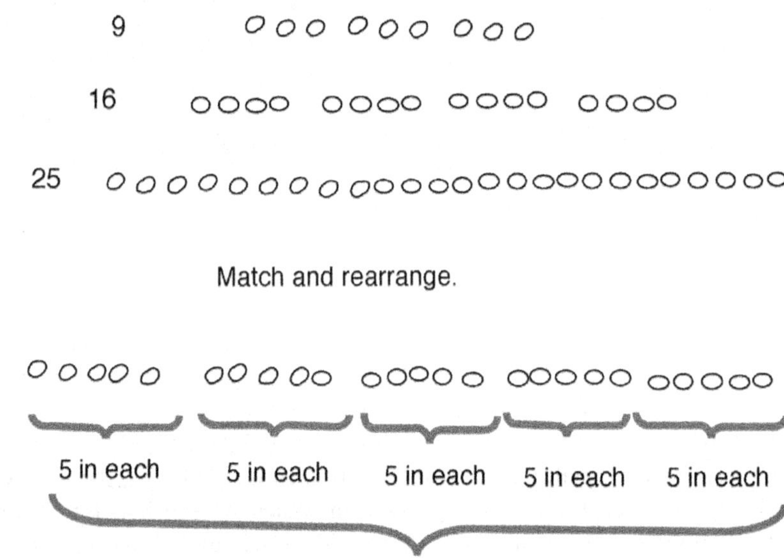

Figure 10.5   Regina's work.

Another middle-school student named John took an approximation route to try to figure out what and how the Pythagoreans must have been thinking. He was particularly intrigued that these early Pythagorean mathematicians, known as mathematikoi,[4] did not have a graphing calculator to help them with the square-root calculations. Instead, the Pythagorean mathematikoi used a strategy called a "method of exhaustion" to calculate the unending decimals. Exhaustion is likely a sentiment shared by many middle-school students, at least before the invention of the calculator.

John's first step was to make a table that included the positive integers and their squared calculations. He chose to initially use an isosceles right triangle to explore the Pythagoreans' thinking about finding patterns. Just adding the two squares certainly did not work to generate the long hypotenuse side, so he reasoned, pretending to be a Pythagorean mathematikoi, that if the sides are squared then the hypotenuse must also be squared, *and* that the actual triangle side length is not the same as the squared measure of that side. This recognition guaranteed that he would *not* be making the erroneous assumption that if $a^2+b^2=c^2$ then $a+b=c$ ever again!

He started his investigation using the case when both legs measured three units for lengths. He recognized that the longer side (hypotenuse) had to be greater than "3" in order to make sure that it was *not* an equilateral triangle, but he knew that it also had to be less than "6." Can you guess why he said that the third side had to be less than six? His notations in Figure 10.6 made sense to him as he worked through his investigation. As long as the results are reasonable and the calculations are accurate, you do not have to understand every notation during your child's (or John's) thinking stages. But if they are willing and able to explain it to you, all the better!

John's exploration was not speedy because he was strategically guessing, checking, and evaluating. Figure 10.6 shows only the first and last pages of his data-generating efforts. He eventually dropped the isosceles-right-triangle idea, and decided to use different lengths for the two leg lengths in his triangles. One of his tested triangles used the 4.24 number from the hypotenuse from his first test and applied it to one of the legs of a new triangle. By now, he had decided that he needed the calculator to save time (and avoid that exhaustion) while he still guessed different numbers for leg measures.

After several more guesses and tests, he decided to try 3.75 as the second leg of his new right triangle. Still trying to limit his use of the calculator (after all, he was trying to think like a mathematikoi), he approximated that the hypotenuse this new triangle would be between 5.6 and 5.7 units. Mathematical thinking is not done at Mach speed, even on a clear day! John may not have found ten thousand ways for it to not work, like Edison purportedly did with the light bulb, but he did establish in his head what was required for the relationship *to work*.

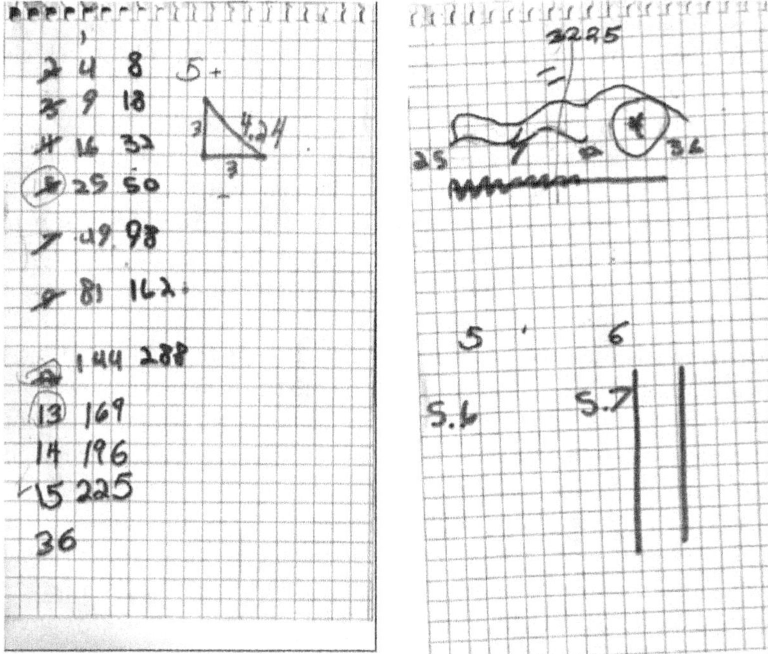

Figure 10.6 John's mathematikoi effort.

Rearranging these numbers into a long line instead of a square area, or collecting and estimating data without using a calculator (or minimize its use), are examples of how Regina and John used flexibility and imagination to make sense of math. Regina, John, and your child too, can join the Chinese, the Egyptians, the Pythagorean mathematikoi, and the other Math Aficionados when thinking about and calculating side lengths of right triangles. When you do, be forewarned that irrational numbers may appear if the right triangle sides are not Pythagorean triples. Not to worry; it took centuries post-Pythagoreans for irrational numbers to be recognized.

TRIGONOMETRY: A RELATIONSHIP BETWEEN SIDES AND ANGLES

Whether measuring the distance to the moon or the distance to a very high tree top, right triangles take top billing. Before trigonometry, you could only figure out missing angle measures in a triangle if you had at least two angles, because you knew that the total had to be 180° for all three angles in a given triangle. Now, with trigonometry and trig ratios, you can use two *sides* of a right triangle in a ratio and figure out *the related angle.*

In the elementary grades, most geometric relationships in shapes are explored separately, using only the sides of shapes or just using the angles of shapes. The relationships that include angles mixed with sides to help with calculations don't happen until the right-triangle study called trigonometry. While trig is based on the right triangle, the applications are not limited to right triangles. "Yet trigonometry, belying its much too modest name, now goes far beyond the measurement of triangles. . . . it has paved the way for the analysis of anything that repeats, from ocean waves to brain waves. It's the key to the mathematics of cycles."[5]

The origin of the word "trig" could be from Middle English *trigg* or from Scandinavian or Norse origins with *trigger* or possibly Olde English *treowe*. Even the early Greeks had a hand in the name with "measuring triangles" as their trigonometria, reasonably using the prefix "tri" for triangles. The use of trig ratios for measuring triangles, however, dates to the Rhind Papyrus and Babylonian work. No surprise there, since many of the original mathematical relationships also had early origins.

Hipparchus of Nicaea (from Chapter 5 and 360° circles) generated the first trig table, thus earning the title "Father of Trigonometry," and was followed by Aryabhata, one of the pi contributors mentioned in Chapter 8 and who was one of the early mathematicians to use the sine ratio as a half-chord. Chords, you might remember, are those one-dimensional line segments found in circles and described in Chapter 4, and Hypparchus' sine ratio was dependent on that chord. Again, this illustrates how mathematicians from different centuries continued to stand on one another's shoulders—an example of how communication in math requires consistency and clarity.

How does knowing about trig ratios with triangles help you help your child with today's homework? The short answer is: because it can provide another context for ratios and similar triangles. The longer answer is: because it can provide quite a few

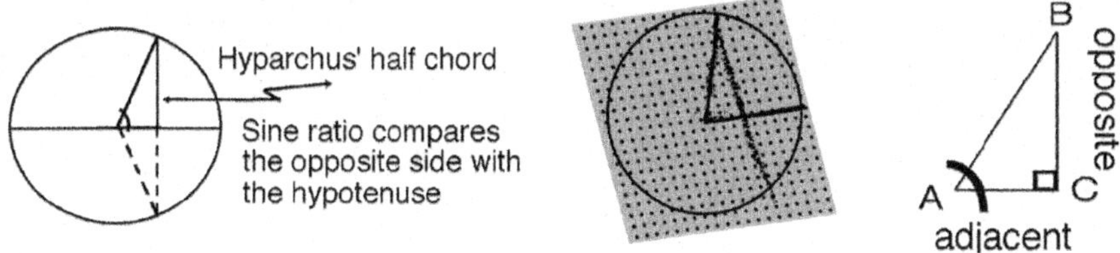

Figure 10.7 Hyparchus and trig arrangements.

answers to many of those "why" questions that either you asked during your math studies or that your child is now asking. Your child learned about ratios as comparisons of measures (not just fractions as part to whole comparisons) in upper elementary school or in middle school. Trig is all about the comparisons of sides, and the sides of triangles are clearly not fractions (parts) of each other.

An assembly with a pegboard, some nonstretchable yarn pieces, and a plumb line helps both visual-spatial and auditory learners to manipulate triangles as the yarn-radius is rotated around the circle. The longer leg of the triangle in Figure 10.7 hangs straight down like a plumb line, the same kind of plumb line that plumbers, carpenters, and fishermen use so that gravity can pull the bob at the end of the plumb line straight down vertically. This is the same plumb-line idea from Chapter 2 that has a Latin vocabulary ancestor in "perpendiculum." A different-colored yarn is stretched horizontally from the center to the edge of the circle, forming a right triangle.

The first three (of the total six) trig functions are sine (abbreviated sin), cosine (cos), and tangent (tan). These names may have come from Latin terms *sinus, complimenti sinus,* and *tangens* or from Sanskrit, or even from Arabic. The last three trig functions represent the reciprocals of the first three functions: cotangent (abbreviated cot) is the reciprocal of tangent; secant (abbreviated sec) is the reciprocal of cosine; and cosecant (abbreviated csc) is the reciprocal of sine. All are related to ratios of side lengths in a right triangle. Reminder: When there is a small box in the corner of a triangle (Figure 10.7), then the triangle is a right triangle.

To identify the ratios in a right triangle, your child needs to choose the reference angle so that the sides have an anchor to refer to the opposite and adjacent locations (Figure 10.7). For example, the sine function ratio is written with opposite leg to an angle (cannot be the right angle) compared to the hypotenuse side (opposite/hypotenuse). Without an identified reference angle, you don't know which side is opposite which angle. Typically, the angle on the base of the triangle is the reference angle; however, that is not a requirement. The only angle in the triangle that is off limits is the 90° angle. The other trig function ratios are included in the glossary.

For the game-enthused or the card deck fanciers, all of the six function equations in trig can be matched together, (see Figure 10.8) each to its specific graph and coordinate pair, challenging all players to the task of winning the game by being the first to score the most points. For the Math Aficionados, or the especially curious player, the amplitude and period playing cards can be included in the matching for a higher winning score.

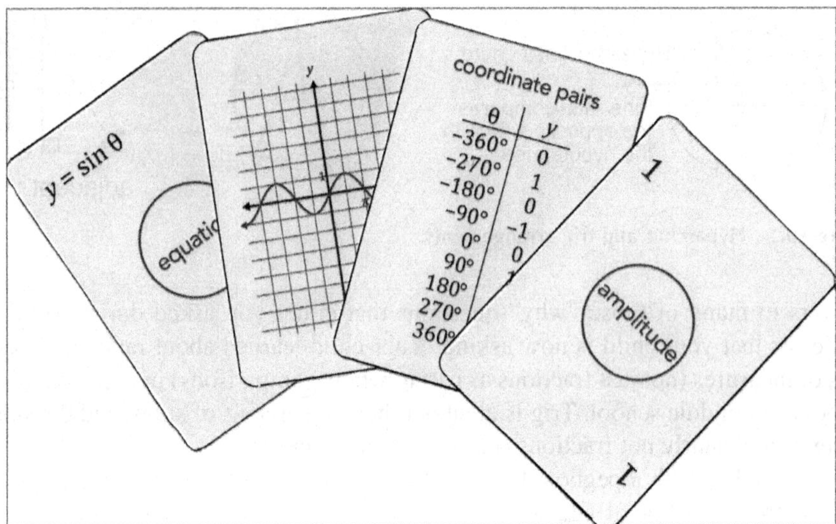

Figure 10.8   Algebra game trig matching cards.

Now comes the similar triangle part. The ratios remain the same no matter how large the right triangle is, as long as the larger or smaller triangle is similar to the original triangle. The angles never change in similar triangles, so the shape itself may get smaller or larger (shrink or grow), but the sides must stay in proportion (there are the equivalent ratios again!) The A, B, and C letters in Figure 10.9 are augmented with an accent mark so that the reader can see the three drawings as being different, but related.

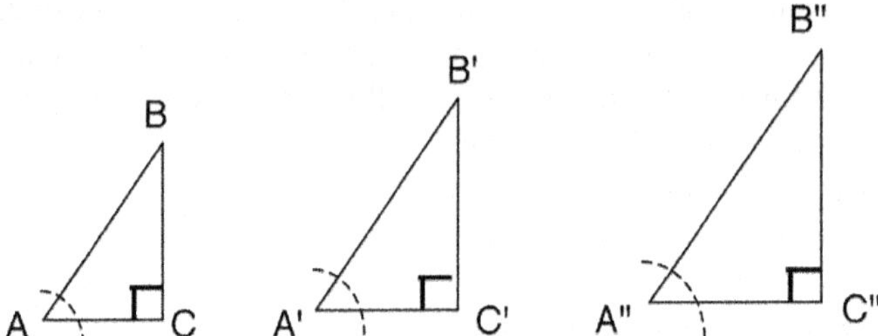

Figure 10.9   Similar triangles share the same trig ratios.

Do you recognize the triangles from the trees, poles, and shadows in this chapter? These triangles are diagrams, freed from the actual slippery poles, rough bark trees, deteriorating surfaces, and elongating shadows. All of the compared ratios in these triangles, when applied to the same orientation of sides, are equivalent. For example, all sine ratios for 30°-60°-90° triangles have equivalent ratios just as all of the sides of these similar 30°-60°-90° triangles are proportional. Here we are again, connecting the trig ratios with triangles with the Pythagoreans.

All cosine ratios for 30°-60°-90° triangles also have equivalent ratios, as do all tangent ratios for 30°-60°-90° triangles, because the triangles are similar triangles. This relationship that keeps the sides proportional is directly connected to the equivalent fractions concept that your child learned in the third or fourth grade. The four triangles in Figure 10.10 look as if they are the same size. However, the number labels for the sides indicate that the triangles are *not* the same size, *but they are similar* and they share the same ratios for sine, cosine, and tangent. The four triangles are not drawn to scale, and serve as a reminder to always check any number labels.

The Math Aficionados will recognize the 30°-60°-90° triangle relationship among the sides from studying the Pythagorean theorem. The rest will recognize some of

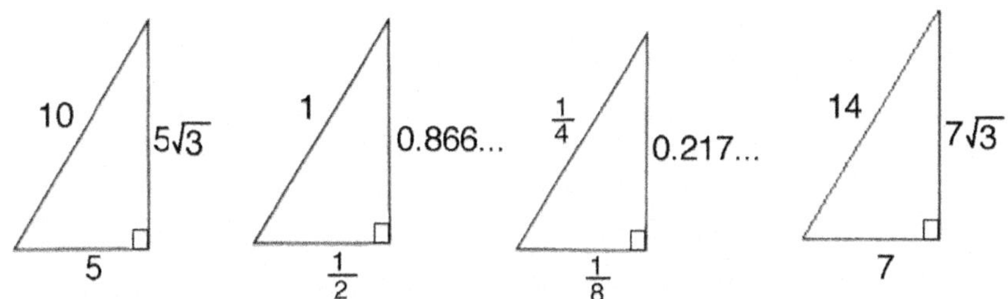

Figure 10.10   Trig relationship of sides meets Pythagorean 30°, 60°, 90° special relationship.

the common characteristics between the hypotenuse and the smaller leg in all three triangles. The 45°-45°-90° triangle relationship is reminiscent of the isosceles triangles from Chapter 1, this time within a trig setting. Make up your own numbers for legs of the triangle, use the Pythagorean theorem and a calculator, and calculate the length of the hypotenuse. There is more to explore; however, the objective is to provide a running start so that your child's imagination is piqued at the right time.

KEEP IN MIND

Understanding how influential the right triangle is, it's not surprising that Johannes Kepler compared the Pythagorean theorem with a measure of gold treasure. As Pappas reminds us, you will see mathematics all around you in right triangles formed by ladders placed against a building, pyramid (tree and pole) shadows, floral arrangements, both sides of a line of symmetry (and height) in equilateral triangles, and many furniture and architectural designs. In Chapter 11, this right triangle continues to spread its influence to the circle (or is it the circle that has influence on the right triangle?)

NOTES

1. Kasner and Newman, *Mathematics and the Imagination*, 144.
2. http://www.goodreads.com/author/quotes/7396.Johannes_Kepler
3. Doczi, *The Power of Limits*, 13.
4. Wertheim, *Pythagoras' Trousers*, 24.
5. Strogatz, *Joy of X*, 115.

Chapter 11

A Circle's Nobility

Bryson [of Heraclea] declared the circle to be greater than all inscribed, and less than all circumscribed, polygons.[1]

—Themistius

Who could have imagined how significant a circle could be to geometry? Archimedes followed Bryson's lead two centuries later as he closed in on a circle's pi ratio by inscribing and circumscribing polygons with circles. The geometry topics in this chapter join Archimedes and other mathematician's efforts to build on Bryson's idea about this rather unassuming circle shape that has such a high rank among so many geometric measures and relationships. Bryson, Themistus, and Archimedes are just three of the mathematicians that set the "circle" stage for another mathematician and creative author to assign the circle's nobility status in *Flatland*.

The circle that the Flatlander could see was only one part of the visitor from Spaceland Voice's spherical anatomy that intersected with the Flatlander's flat plane surface. For now, the circle relationships in this chapter will only describe relationships on that flat plane surface, and invite an adventure into Spaceland's spheres and relationships with polyhedra when your child is ready. That visitor from Spaceland spoke to the Flatlander about having "many Circles in one," so we can imagine those many circles revolving around the visiting Spacelander as a sphere—a three-dimensional circle, so to speak.

By their very nature of being round, circles play a unique role with regular polygons that likely contributed to their Flatland nobility title. The circumscribing circles share diagonals, diameters, and lines of symmetry with polygons. Because circles have many, *many* lines of symmetry (all the way around like the diameters), it makes sense that some of those lines of symmetry and diagonals in polygons would coincide with one or more of the lines of symmetry and diameters of the circumscribing circle.

The lines of symmetry in all of the shapes, including circles, go through the center of each shape. If the polygons are inscribed in circles (polygon vertices touching the circle), then other relationships begin to reveal themselves as shown in the examples in Figure 11.1; namely, overlapping lines of symmetry, diagonals, and

111

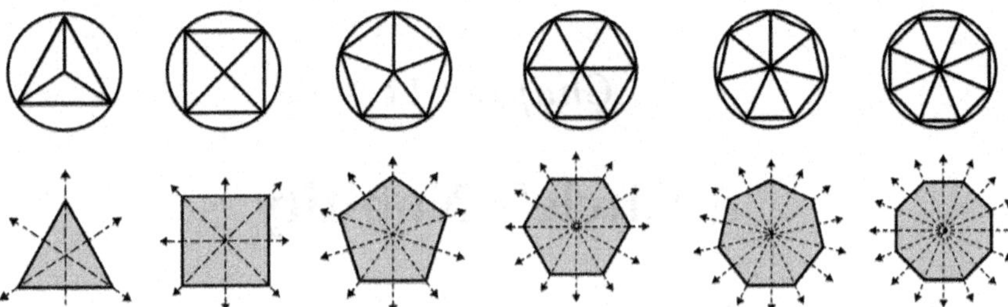

Figure 11.1   Lines of symmetry, diagonals, diameters, apothems - finding overlaps.

diameters. The internal isosceles triangles from Chapter 2 return, the right triangles
with their apothem legs from Chapter 5 come back to be used as heights for the
triangle areas, and angle degrees from Chapter 4 can be seen in the triangle rotations from Chapter 7, all coming together into one image. Do you see the shared
overlaps?

The top row in Figure 11.1 show regular polygons, each one inscribed in a circle,
and all of the polygons have their matching isosceles triangles rotating from the
center. These are the triangles waiting for their related apothem from Chapter 2 so
that their areas can be calculated. The second row contains the same polygons *not
inscribed in a circle*, but each polygon shows all of their lines of symmetry. Do
you see the shared lines of symmetry with diameters of circles? Do you recognize
the diagonals of polygons that share a position with the circle's diameter? Do they
all share or just a few? Get out your tracing paper and your imagination and start
tracing!

Recognizing these shared diagonals, diameters, lines of symmetry, and radii of
circles allows your child to recognize spatial connections within circles and polygons,
and also gives them permission to continue to look for more. For example, that apothem didn't just appear out of nowhere; it came from a line of symmetry when a height
was necessary to calculate the area of the inscribed polygon. Examining the different
connections between the odd-numbered sided polygons and the even-numbered sided
polygons opens more connection doors. The images in Figure 11.1 will get you started
down yet another connections road.

If your creative juices are flowing and your imagination is in high gear, you might
recognize a few of these Figure 11.1 basic designs in hubcaps, quilts, pillows, family
crest and other insignia, art images in paintings, ring settings, stained glass windows,
water main covers in streets, wreathes, and flower arrangements. The list will continue
to grow as more imaginations create new designs.

As long as these regular polygons are inscribed in circles, it would seem reasonable
for the angle degrees in these polygons to have a connection with the total number of
degrees in a circle. Hipparchus gave us the 360° measure at the center of every circle
and every triangle no matter its size has a total of 180°, so if all of the triangles are
added that means all of those triangles with 180° can be multiplied by the number of
triangles in the particular polygon. That's a lot of degrees for a simple polygon! In
fact, it's too many; it is 360° too many. All of the triangles meet at the center, so that
center circle is not part of the polygon's angles.

ANGLES AND TRIANGLES UNDER CIRCLE CONTROL

Both Theo and Lizzie benefitted from knowing that angle measurements are maintained when rotated, slid, or reflected with the congruence and symmetric transformations (see Chapter 7) or even "relocated" in a circle. Having Hipparchus bring in the 360° circle allowed Theo, Lizzie, and another student named Claire to begin to connect topics by containing them in circles.[2] Claire was using the polygon pictures in Figure.11.2 to calculate angle degrees in polygons and use the circle's 360° degrees to finalize her formula.

Claire was a freshman in high school and learning about linear equations in her algebra class. When she was given a problem about inscribing polygons in circles in another exploration setting, she started drawing those regular polygons from middle school days. She collected the data for the tables in Figure 11.2 and put her algebra information to work to eventually generate her linear equation for the total number of degrees in all of the triangles and then subtracted the 360° for the center circle. Her final equation was y=180°x-360°. Later, she would draw triangles with polygon diagonals (like the ones shown in Figure 11.8, later in this chapter) so she wrote another equation, y=180° (x-2).

All of those other triangles from Chapters 2 and 3 (still staying in the convex arena) are also contained in the circle, like the examples in Figure 11.2. All of the triangles still contain 180° as the total of the triangle's angle measures; however, the *measures of the angles* in those triangles are connected to the circle's arcs in a very particular way. If the angle is located on the circumference like the ones in Figure 11.3, then its

Regular Polygon Shapes							
Number of inner triangles	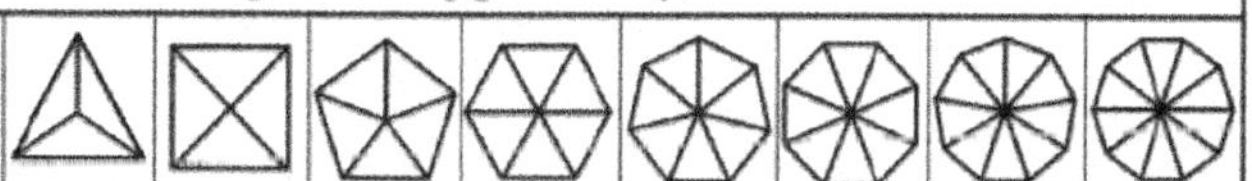 3	4	5	6	7	8	9	10
Total degrees in the polygon	540°	720°	900°	1080°	1260°	1440°	1620°	1800°
Total of angle degrees around the perimeter	180°	360°	540°	720°	900°	1080°	1260°	1440°

Claire's Table of collected data					
Number of sides	3	4	5	...	x
Number of triangles	3	4	5	...	x
Total number of degrees (y)	540	720	900	...	180 x

Claire subtracted the 360° from the center. y = 180 x - 360

Figure 11.2 Claire's work.

measure is one-half of the arc measurement; if the angle is in the center like the ones in Claire's polygons circumscribed as in Figure 11.1 and Figure 11.2, then the center angle measurement equals the arc measurement.

In the spirit of confirming Sophie Germain's assertion in Chapter 12 that "geometry is but figured algebra," the straight-angle from Chapter 2 looks very much like the number line. The swing arm of the Positive and Negative Rotating Number Line (shown in Figure 11.3 for identifying positive and negative integers)[3] looks like the straight angle that totals all three angles in the triangle, regardless of their size, from Chapter 5. In fact, that swing arm on the number line can make those angles that were in Theo's test question in Chapter 1.

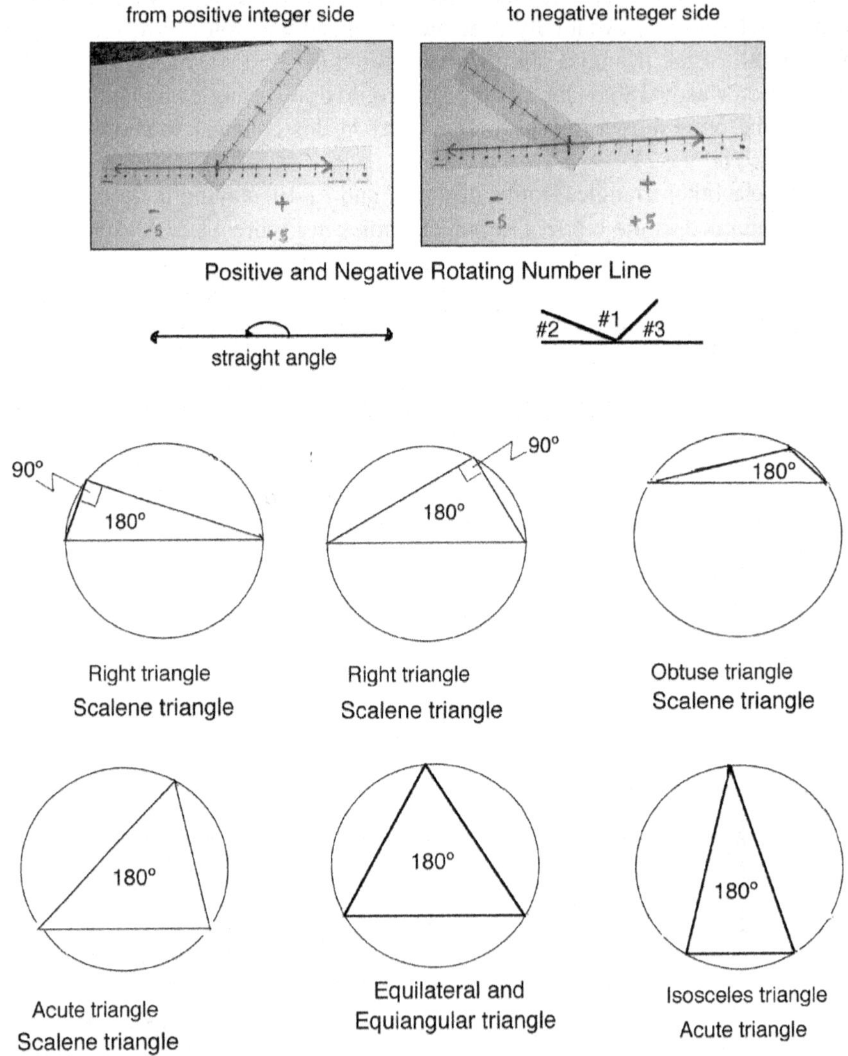

Figure 11.3   Theo's permission to move angles meets circle angle measures in inscribed triangles.

Theo's triangle sum that totals 180° is not an accident. The right triangle that shares its hypotenuse with the circle's diameter also shares its right angle with the semicircle arc; yes, half of Hipparchus' 360° circle. The other two angles in the right triangle *must have* degree measures that would total another 90° so that the angles in the triangle could maintain the 180° total. When those triangles from Chapter 2 are circumscribed by a circle, new questions arise and new insights are open to discovery. Are all polygons circumscribable? Are there any limitations for making a polygon circumscribable? Good questions suggest investigations and confirmations.

Prove It to Yourself:

Find a round coffee filter, or round piece of paper to fold.

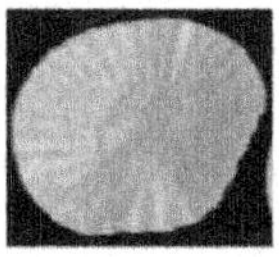

First level: fold the round paper to form a right triangle. Measure the right angle with a protractor for your Pólya test.

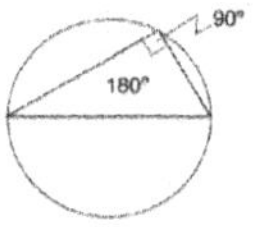

Second level: cut out rhe right triangle and fold all of the angles to the center of the circle.

Third level: fold a round paper to form a 30° - 60° - 90° triangle. There are several ways to do this.

Figure 11.4 Task Box: Prove It to Yourself.

CIRCLE RELATIONSHIPS AND ALGEBRA TRANSFORMATIONS

Another line in Rita's poem addressed the circle area as it could be calculated with this same formula for a polygon, just with some imagination and adjustments. Though Rita and fourth-grader Sally (from Chapter 5) never met, both of them used the idea of using polygons to approximate the area of a circle. Rita saw the polygon sides as getting smaller and smaller so that the perimeter began to approximate the circle circumference. Sally used a different way to close in on the circle area by using her sixteen triangles.

The area of a circle can be found
By inscribing a regular polygon that is almost round.
The apothem approximates the radius,
Don't worry, I have more to address.
The perimeter approximates the circumference.
Now this is the part that gets a little tense
½ ap = ½ r times (2π times r) equals π times r squared.
Gee, I am really scared.

Sally used sixteen right triangles in Chapter 5 to approximate the area of her circle. She divided the circle into four congruent sections and then subdivided each section into four right triangles. By dividing the 360° by 16, and subtracting that result from 90°, Sally calculated the angle measures in her right triangles. Look closely at Sally's project to see why her division worked. The triangle angles have 22.5°-67.5°-90° measures. A good question to ask at any grade level when the area of a circle is the topic could be: "If Sally used right triangles with one 30° angle, how many triangles would she need to cover a circle?" How about other angle measures?

Another rearrangement of a circle's area to generate a rectangle was suggested in Chapter 5 and is included here as a reminder about Germain's assertion that geometry and algebra go together. The first photo in Figure 11.5 shows a leather trim for the circumference "C" of the circle. The circle is cut into wedged slices and pulled apart so that the circumference is now cut into two semicircles. The two semicircles have several wedges so that when the wedges are pushed back together in the third photo, an image of a bumpy rectangle begins to show.

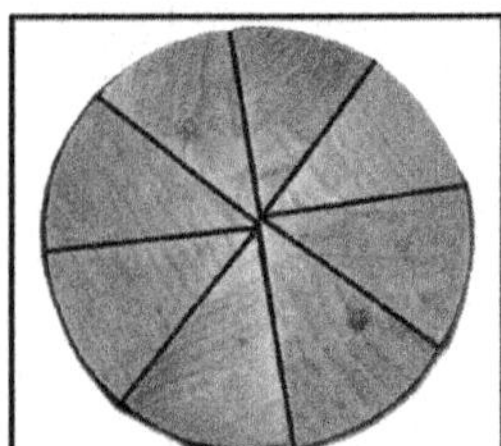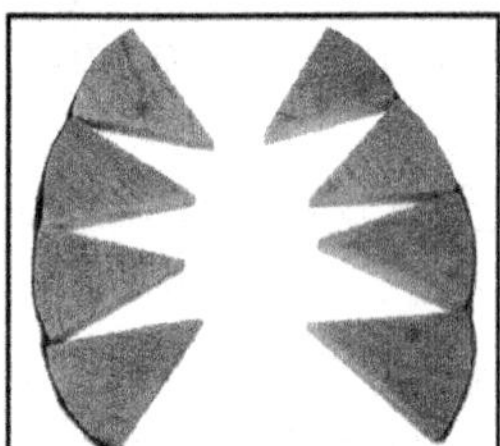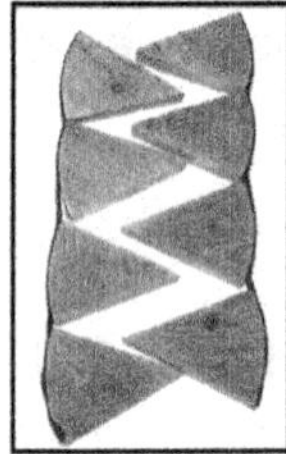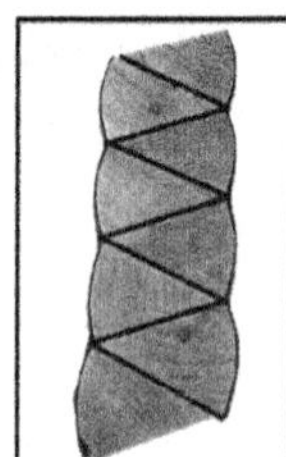

Figure 11.5   Circle conversion to bumpy rectangle.

The radius of the circle becomes the width of that bumpy rectangle and the length of the bumpy rectangle is now one half of the circumference (½C or ½πd). In the last photo, the bumpy rectangle is reassembled so the area can be written as base times height. Start with the rectangle and multiply the length of the rectangle (using circle measure as "½ C") times the base (use "r" for the radius). Rearrange, substitute, and multiply ½C•r (translated as one half of the circumference times the radius because the "dot" replaces the usual x symbol for "times") to ½π•d•r (translates to one half times pi times the diameter times the radius).

Continue rearranging, substituting, and multiplying to end with ½π2r•r to show the completed algebra transition of the area of that original circle to $\pi \cdot r^2$, the area of the circle that we started with. The transition evolved from base times height (some still say length times width, which is also correct) symbolized with (½C or ½πd) to $\pi \cdot r^2$, which is the area of the circle. A reminder from Chapter 8: the circumference is a little over three times the diameter. This algebra process is still about composing and decomposing the circle, but this time using symbols.

Converting the area of a circle into a very close approximation of a rectangle's area uses a mathematical idea known as "limits," and as Strogatz describes it, "when you finally get to the wall [rephrased as a limit], it becomes simple and beautiful, and everything becomes clear"[4] when that bumpy rectangle eventually represents a circle's area. How many circle wedges would it take? Archimedes supposedly used ninety-six sides of polygons to approximate pi, so imagine ninety-six wedges and use that strategy called "method of exhaustion."

RETURN OF THE APOTHEMS FOR POLYGON AREAS

Right triangles appear as part of all of those regular polygons: equilateral triangle, square, pentagon, hexagon, and the rest of those other n-gons. All of these regular shapes have a right triangle (shown in Figure 11.6) inside the polygon. The right triangle height mentioned in Chapter 4 in this central location keeps the "apothem" name. This apothem is one of the legs of the right triangle *and* has the same length as the radius of the inscribed circle of the polygon (see Figure 11.6). The hypotenuse of this right triangle also serves as a radius, the same radius of the circumscribing circle *around the regular polygon.*

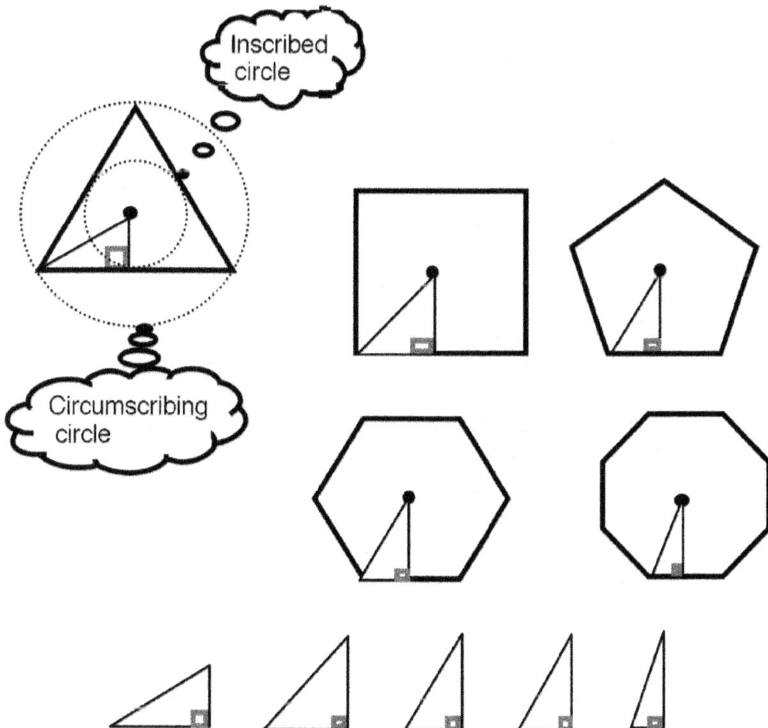

Figure 11.6 Apothems return for more areas.

When all of these triangles are removed and redrawn separate from the regular polygons, they are still right triangles (remember that the small corner square box means that the 90° angle is guaranteed). The right triangles from the triangle and hexagon are congruent and share a special Pythagorean relationship that you read about in Chapter 10; the right triangle from the square also shares a special, but different, Pythagorean relationship with the octagon and is also suggested in Chapter 10 to pique your child's imagination.

A hint about a formula to find the area of a regular polygon that includes using the apothem as a triangle height was provided in Chapter 5. That formula is A=(a•p)/2, letting "a" represent the apothem of that polygon and "p" stand for the perimeter of the polygon. With flexibility, rearranging, *and* the commutative property for multiplication, the perimeter of the polygon can be separated into all of those shorter bases of the triangles in a polygon. By rearranging parts in the formula, the revised formula expression is a matter of adding all of the areas of the internal triangles. "($\frac{1}{2}$b$_1$•a)+($\frac{1}{2}$b$_2$•a)+($\frac{1}{2}$b$_3$•a)+($\frac{1}{2}$b$_4$•a)+ . . . the rest of the triangles."

Just as Sally (from Chapter 5) used triangles to approximate the area of her circle in the fourth grade, these polygons with apothems are also used to approximate areas of circles. How many triangles would be necessary? Archimedes used ninety-six sides of imagined polygons, so ninety-six sides would mean ninety-six triangles, each with its own apothem. Therefore, before giving up, try imagining ninety-six isosceles triangles in the polygon formula for area. Use a calculator or your own version of a method of exhaustion to calculate the areas for all ninety-six triangles.

Prove It to Yourself:

Draw the regular polygon shapes with a right triangle in the central position. Make sure that all of the polygons have the same radius in each circumscribing circle. (See figure 11.6).

Trace the triangles and try to find which ones match exactly. You may rotate them to test the matches.

Some triangle examples for the 180° are: 90°+60°+30° and 90°+45°+45° and 90°+54°+36° and 90°+67.5°+22.5° and others for regular polygons with more sides. Want to venture a Bombellian wild thought guess about which angle combination goes with which right triangle picture?

Answers:
Square has 45,45,90. Triangle and hexagon have 30, 60, 90.
Pentagon has 36, 54, 90. Octagon has 22.5, 67.5, 90.

Figure 11.7   Task Box: Prove It to Yourself.

DIAMETERS, DIAGONALS, AND RATIOS REVISITED

When your child places the rolled-out tracks of the shapes in Chapter 4 next to their traced diagonals and diameters (on tracing paper) shown in Figure 11.8, they can see how those one-dimensional line segments have another connection with sides of triangles that were circumscribed in Figure 11.1. The pentagon would make isosceles triangles with two congruent sides. A hexagon would make a different isosceles triangle and share a diagonal with the circle's diameter. These polygon diagonals (and circle diameter) also share a relationship with their polygon's perimeter (and circle's circumference).

More algebra and more drawing will confirm more examples of Germain's idea about relationships between figured algebra and written geometry. The circle's circumference relationship to its diameter is the pi ratio. What do you think a relationship might be between a regular hexagon's perimeter, or side lengths, and its diagonals? Between a regular octagon's perimeter, or side lengths, and its diagonals? After the square and pentagon, the hexagon, octagon, and other polygons have more than one length for their diagonals. Now Archimedes' work with regular polygons to calculate pi probably seems more reasonable.

All of the diagonals in figure 11.8 are drawn from only one vertex so that the relationships may be easier to recognize. For example, a pentagon actually has *ten* diagonals, two diagonals extend from each vertex. Drawing all of them might make it difficult to distinguish the diagonal lengths. However, refer to Figure 4.4 in Chapter 4

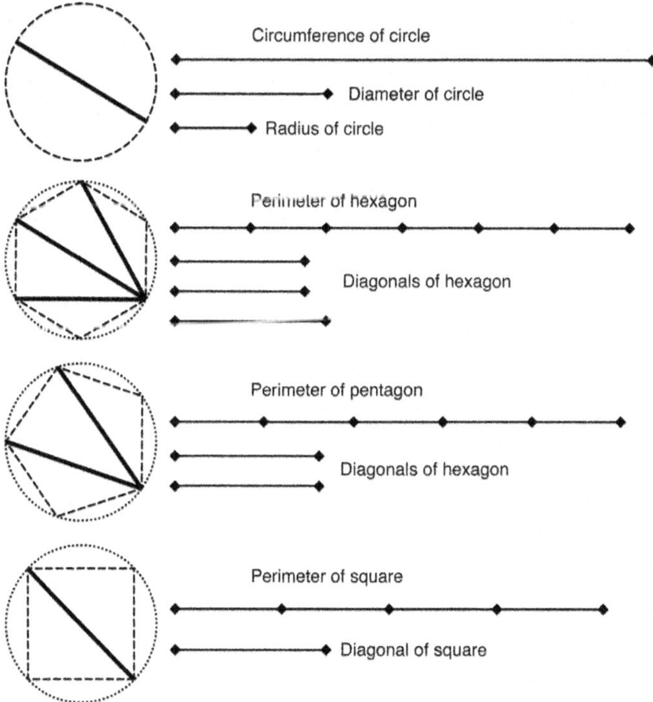

Figure 11.8   Diagonal and diameter relationships.

to see all of the diagonals for the square, pentagon, and hexagon. The rolled lengths for each perimeter and side lengths show an increasing trend. Calculating these diagonal lengths requires some imagination, angle measures, trig, and the Pythagorean theorem.

All of the circumscribing circles are congruent, so, if you wish, you may assign a diameter length for the circles. Rotations, triangle height-finding, isosceles-triangle relationships, and some algebra assistance from Claire can help you uncover some of the diagonal lengths. And if all else fails, trig ratios can give you that extra boost you might need. Start asking questions, even questions that you do not know an answer to yet, as you begin to tackle these diagonal, diameter, and perimeter relationships.

Asking a good question means that you don't have to know the answer, or any of the possible answers or solutions, until after some exploration. Halmos, Pólya, and Poincaré are only three of the many mathematicians who were willing to ask questions without knowing a solution right away. Paul Halmos suggested vague guesses and visualizing broad generalizations before becoming convinced of their truth through proof.[5] George Pólya suggested guessing, strategic guessing and constant evaluation. Henri Poincaré is quoted as asking, "How is an error possible in mathematics?"[6] All encouraged guessing, insight, and pursuit of mathematical truth.

KEEP IN MIND

Circles can help many of the geometry measurements and elements make more sense because they actually contain these measures and show how the elements are related to each other and to the circle. When a circle circumscribes those triangles from Chapter 2, new insights are available for discovery. Are all polygons circumscribable? What requirements make a polygon circumscribable? Asking questions can initiate new ideas and new ideas can help both you and your child think mathematically "out of the traditional box" to find some new and different ideas and connections.

NOTES

1. Heath, *A History of Greek Mathematics*, 224.
2. Draper, *Winning the Math Homework Challenge*, 81.
3. Draper, *User-Friendly Math for Parents*, 62.
4. Strogatz, *Joy of X*, 126.
5. Halmos, "Mathematics as a Creative Art," http://www-history.mcs.st-andrews.ac.uk/Extras/Creative_art.html.
6. Poincaré, *Braineyquotes.com*.

From Figures to Scale Models, Then on to Calculus Connections

Algebra is but written geometry and geometry is but figured algebra.[1]

—Auguste Antoine Le Blanc (the pen name for Sophie Germain)

It sounds like written geometry and figured algebra are inseparable. The connection between algebra and geometry *is* the mathematics! Chapter 11 described several examples of how one assists the other, now in Chapter 12, there's more! The two hemispheres of the brain (oversimplified for this discussion to assign the right hemisphere to geometry and left hemisphere for algebra symbols) "talk to each other" over the three hundred million nerve fibers in the corpus callosum bridge. Shapes and diagrams can enhance visual understanding while symbols and vocabulary explain the sequential logic behind the understanding.

Both hemispheres are necessary for mathematics and using the whole brain can ease the anxiety around math. Once your child recognizes that geometry and algebra can be a "math team," then their experience in math is forever changed, especially for the visual learner. The written part of the geometry of shape relationships connects to the idea of algebra and the arithmetic associated with calculations around images. These two concepts of written geometry and figured algebra work together as the topics in this chapter will show by addressing the full spectrum from first grade through to a gentle introduction and connection to calculus.

SHAPES AS FIGURED NUMBERS AND ALGEBRA

Shapes can provide the visual relationships and let us explore and explain all kinds of patterns among numbers, starting with the number line in first grade. That number line is a straight-line shape in the first dimension and shows a visual model of how *our numbers are ordered*. Place value categories for grouping numbers also uses dimensions to illustrate how *our numbers are organized*. Those two essential organizations for numbers *depend on diagrams and dimensional shapes* to illustrate number relationships.

In the first grade, the number line shows how the numbers are ordered, staying in the first dimension. Sometime around the fourth grade, that line represents a straight angle, a concept that marks the beginning adventure into the second dimension. Later in the middle-school years, that one-dimensional line is duplicated and rotated to define two-dimensional locations in a coordinate system. After that, another duplication of the number line allows us to represent the third dimension with a three-dimensional coordinate grid. That simple, seemingly innocuous straight line has a grand, expansive organizational impact throughout mathematics.

Return to the first grade for that other shape-dimension connection that helps define the place value organization for grouping numbers into tens categories. Three shapes in three dimensions represent these groups to show how dimensions play their part. The first group of ten is shown as a ten-stick that looks a lot like a line for the first dimension. The second grouping for ten puts ten of the ten-sticks together to make a hundred that makes a shape that looks a lot like a two-dimensional square. Stacking ten of those square-shaped hundreds generates a cube, a three-dimensional shaped polyhedra, to represent that next thousand category in place value.

Those decompositions from Chapter 1 and Chapter 5 show more examples of Germain's figured algebra through formulas. The trapezoid in Figure 12.1 shows how the formula for calculating the area of a trapezoid, "$A=1/2 (h)(B_1+B_2)$" (the "h" stands for the height of the trapezoid and the two bases, B_1 and B_2) is another kind of shape decomposition. With a little rearrangement and a lot of flexibility, *and* the commutative property of multiplication, the "1/2" is multiplied by "(B_1+B_2)" to *find the average of the two base measures.* Voila, the average of the trapezoid's two base measures is multiplied by the same height to calculate the area!

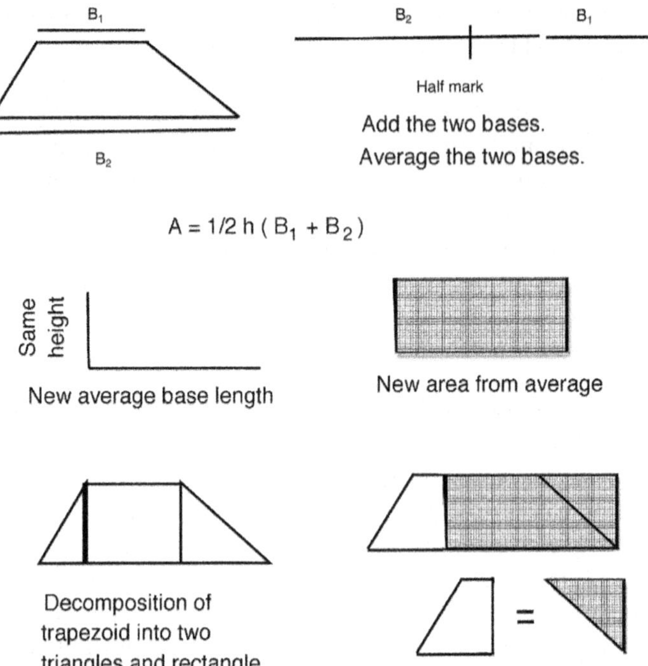

Figure 12.1   Figured algebra area with trapezoid decomposition example.

The gray area in the new "averaged base" rectangle in the middle row of Figure 12.1 covers most of the original trapezoid with some extra gray space left over; that gray space is external to the original trapezoid but will fit into the not-yet-covered white piece of the original trapezoid. That extra gray space in the last row has the same areas as the "extra noncovered part of the original trapezoid." The formula calculation presents another type of decomposition that is the written geometry of the two-triangle-and-square selected decomposition figured algebra shape. You and your child may choose which one makes the most sense to you.

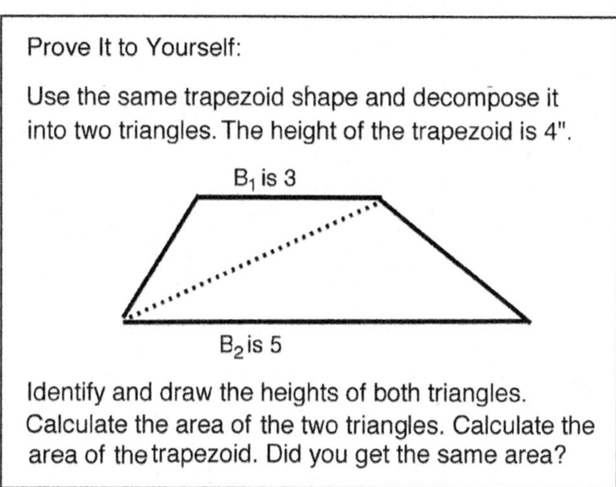

Figure 12.2   Task Box: Prove It to Yourself.

The transition of the circle shape to a rectangular shape in Chapter 11 is another example of what Germain refers to as figured algebra and written geometry. The written algebra symbols for the rectangle area as "length times width" were translated into the symbols for a circle in geometry. The sequence in the transition of *how the geometry of a circle shape could morph into a rectangular shape* happened when the geometry terms were expressed in rectangular terms, both showing the same area. Like the trapezoid and circle examples, and the area of polygons also in Chapter 11, geometric decompositions are expressed both symbolically and geometrically.

WRITTEN GEOMETRY THROUGH KEEPING RECORDS

When shapes change, the relationships between the shapes also change. In Chapter 9, the sides of similar shapes doubled or tripled and changed their areas, creating a new and different relationship between their areas. What do you suppose might happen to areas and perimeters if your child increased *only one pair* of sides of a shape but not the other sides? Drawing shapes and writing the number data for each allows your child to witness the changes as they happen. Keeping records isn't a new idea; it can begin as early as your child is able to keep score in a game. A pattern begins to emerge as the numbers in a record reflect these shape changes.

The rectangle shapes and tables in Figure 12.3 show examples of how the grow-in-only-one-direction patterns affect the areas and the perimeters of rectangles.

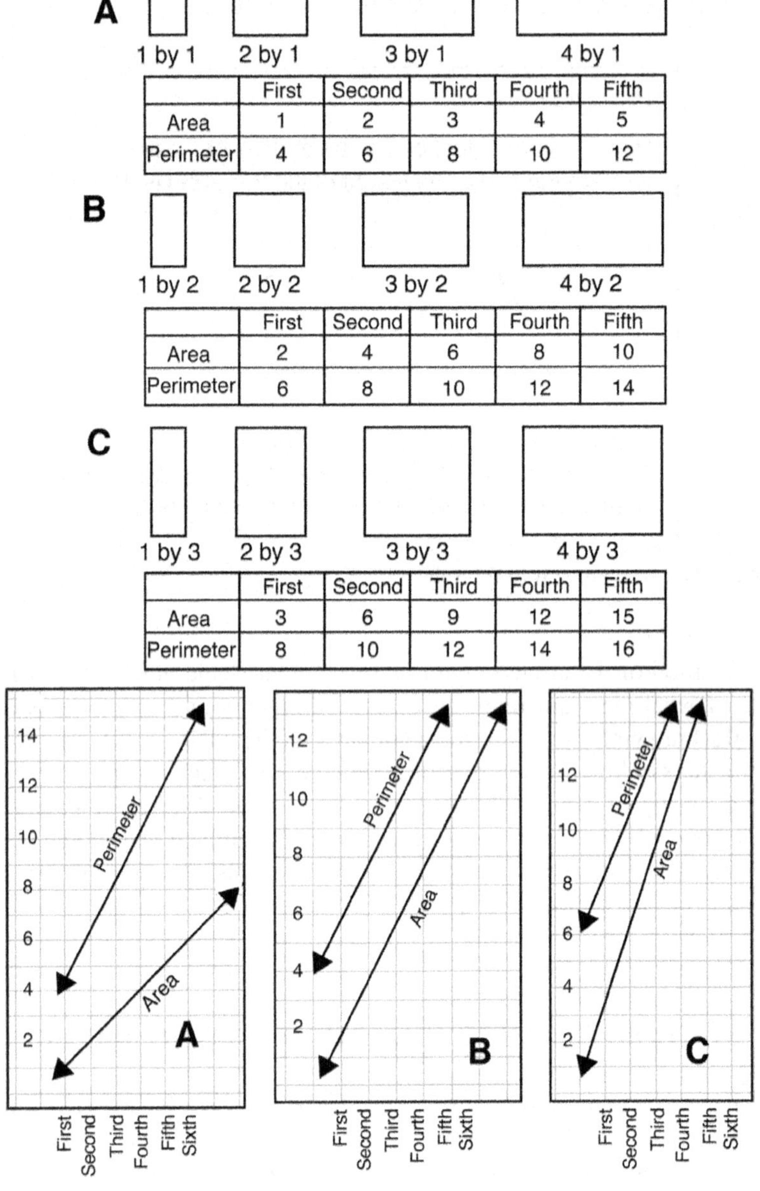

A

1 by 1   2 by 1   3 by 1   4 by 1

	First	Second	Third	Fourth	Fifth
Area	1	2	3	4	5
Perimeter	4	6	8	10	12

B

1 by 2   2 by 2   3 by 2   4 by 2

	First	Second	Third	Fourth	Fifth
Area	2	4	6	8	10
Perimeter	6	8	10	12	14

C

1 by 3   2 by 3   3 by 3   4 by 3

	First	Second	Third	Fourth	Fifth
Area	3	6	9	12	15
Perimeter	8	10	12	14	16

Figure 12.3 Areas, perimeters, and tables, connected to graphs. The visual trends are in the graphed lines.

The records show how the perimeters change and also how the areas change *in a different pattern*. Perimeters apply to one-dimensional shape measure, so those measures can be compared with each other. Areas are a two-dimensional shape measure, so they are compared with each other. When the two kinds of comparisons are each shown as a graph, then *their respective trends* can be compared across the dimensions.

The three graph pairs for these three rectangle trends show how the areas and perimeters can be compared. The rectangles that have a constant side measure of two show parallel line trends; the other two pairs of line comparisons are not parallel. Look at the tables recording the areas and perimeters of these rectangles for the number pattern that indicates these graph patterns. Some good questions to explore are: "Would one rectangle ever have the same integer number for perimeter as it has for area even though the units are different?" and "What would the graphed lines look like if that should happen?"

Although the recorded areas and perimeters in these tables are positive integers (rectangles don't have negative areas), there are more rectangles in between the ones in these tables. When a graphed line is not broken, that means that all of the fractional numbers in between the integers will make rectangles, too. For example, a rectangle in table A could have side measures of 2.5 by 1 that would be inserted in the table between the second and third entries with a 2.5 for area and a 7 for the perimeter. The graphed line already goes through the 7 on the perimeter line and the 2.5 is already on the area line graph.

The graphed lines in A appear to never cross as the lines increase. The two lines for area and perimeter in B are parallel. What do you think would happen to the graphed lines for perimeter and areas in the line graphs in the C tables for rectangles that have dimensions of 5x3, 6x3, and 7x3, and others? What will the graphed lines for area and perimeter look like in graphs B for the rectangles with dimensions 5x2, 6x2, and 7x2, and others? If you are feeling like George Pólya, what do you *think* might happen? Or if Paul Halmos appeals to you, how would you establish the truth of your ideas? Draw more rectangles and add to your table records.

If two-dimensional areas are predictable, should three-dimensional solids also have predictable volumes? As long as your child can control the way the sides are changed, then there is a possible predictability ready to be unearthed. It is always easier to start with changing only one piece at a time. So for the two examples of solids in Figure 12.4, only one dimension has been changed on each solid. How does the volume change for each of the changes? How many single cylinders are in the doubled-length cylinder? In the tripled-height cylinder? How many of the single rectangular prisms (shoeboxes) can your child see in the doubled-sided longer prism?

Changing only one measurement at a time makes it easier to recognize trends and patterns. Changing two or more measurements *at the same rate* presents a different challenge but is still predictable, like the shapes explored in Chapter 9. Changing measurements at different rates will change the solid altogether, and discovering any possible trends and patterns will involve those spatial visualization skills.

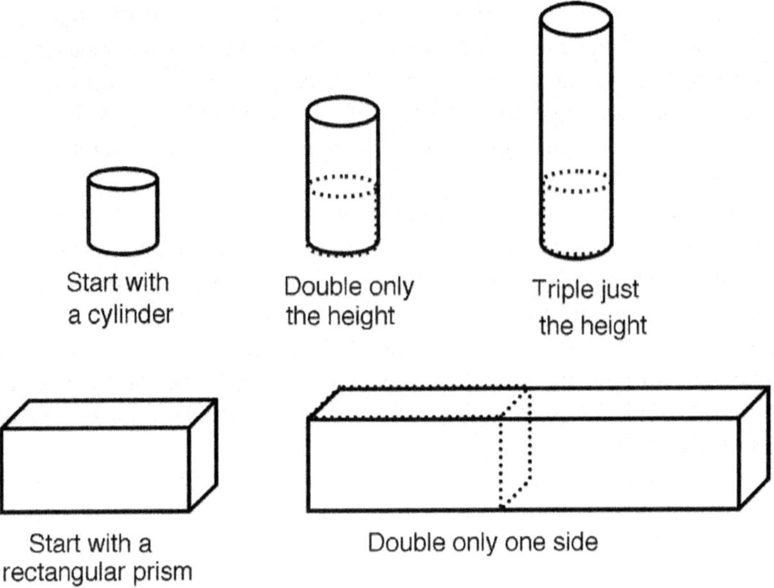

Figure 12.4   Change one measure to predict volume change.

TRIANGULATION, SCALE MODELS, AND SIMILAR TRIANGLES

Seventh-grader Lizzie was initially disturbed about how an angle could be rotated and moved about while still maintaining its original measure. Being able to see this constancy of measurement is very helpful for any calculations, especially the ones with triangles that are not fixed on pages in books. Surveyors make use of these "mobile" triangles when they calculate inconvenient distances such as the widths of a lake (see Figure 12.5) or the heights of mountains (or pyramids or flagpoles) using triangulation, a strategy that uses similar triangles to calculate inaccessible unknown distances.

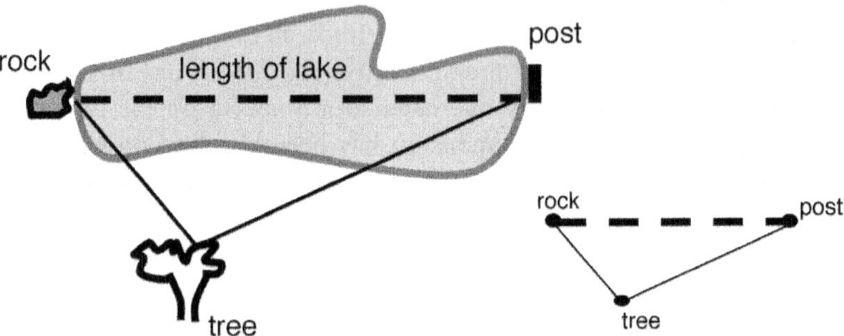

Figure 12.5   Triangulation with similar triangles

The most appropriate strategy for triangulating solutions will depend on the available measurements. Proportions apply to all similar figures and the Pythagorean theorem applies to similar right triangles. Trigonometry ratios apply to similar right triangles or, if the measurements allow, to similar triangles that can be decomposed into right triangles by using the law of sines or law of cosines. The surveyor needs to be familiar with all strategies in order to make the most reasonable decision.

Architects also use the idea of similar proportions when they build scale models for their building projects. Toy manufacturers create scale models of dollhouses, trains, cars, and airplanes, and chefs bake scale model gingerbread houses out of dough. Those ratios from the sixth grade are put to practical use with the one- and two-dimensional shapes and measurements from grade four to produce these replicas later.

Your child may not yet be an architect or toy designer, but she does have creativity and imagination at the ready. Heather made some cardboard eyeglasses in fifth grade, proportionally sized to fit a great white whale, to learn how proportions can make very large measures into more manageable calculations. (Heather also calculated how many baby elephants it would take to "fill a whale.") Later in the eighth grade, she "had another glorious week," like young Dixon's fourth-grade colonial village in Chapter 1, when she built a scale model of the Great Wall of China (Figure 12.6), complete with towers and scaled areas of the surrounding provinces.

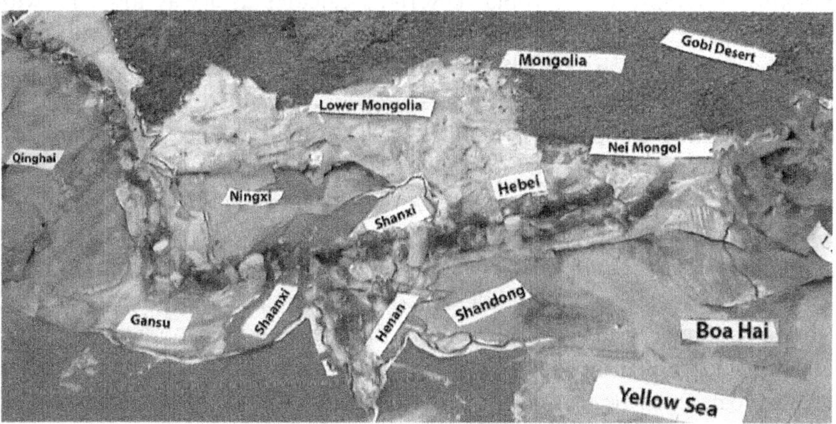

Figure 12.6   Heather's Great Wall of China scale model

Two eighth-grade twentieth-century "ancient surveyors" teamed for their version of a "glorious week" to make a scale model of a mountain on the Greek island of Samos. The project required a great deal of geography, data collection, math relationships with similar triangles and the Pythagorean theorem, calculations, language expression, and history—all these multidiscipline efforts were required in order to show how Greek surveyors planned, designed, and successfully tunneled through a mountain from both sides *and met in the middle*, without the aid of modern cranes or AutoCad software, to get water to their village.[2]

A GENTLE CONNECTION TO CALCULUS

The circle wedges in Chapter 11 got thinner and thinner, the number of triangles got skinnier and skinnier, and Archimedes's polygon perimeters got smaller and smaller, all using an idea of constant change that calculus calls a limit. Calculus answers the basic question of how close your child can get to calculating how much he might need for a solution—an area, a perimeter, a maximum inventory, or other totals that are involved with changes.

Two different mathematicians in two different locations achieved the seventeenth-century development of the formalized study, known today as calculus, again without social media technology for collaboration. Their theories were based on the ideas of many earlier mathematical giants contributing to the same concept. The two mathematicians that get most of the credit for this formalized study are Isaac Newton and Gottfried Leibniz. Newton called the idea "fluxions" and Leibniz called it "finding differences," and both wrote about infinitesimals.

Calculus is about studying how one thing changes in relation to another item, and since things in our lives are always changing, it is a reasonable deduction that calculus is the appropriate tool to calculate these changes. Understanding the connections between ratios from sixth grade to the rate of change studied in later years is essential to understanding the changes represented in calculus. Again following Germain's lead, the figured algebra of geometry is explained with written geometry.

Sawyer suggested the strategy of "collecting the [data] evidence [so that] the discoveries make themselves."[3] Tobias described calculus as a way to "cope with and measure the process of change."[4] Strogatz collapsed the calculus idea into this statement: "Roughly speaking, the derivative tells you how fast something is changing; the integral tells you how much it is accumulating."[5]

With these three components—collecting data, figuring out a rate of change, and calculating how much you need to accumulate (like an area) for what you need for your goal—your child is now prepared for a gentle ease into calculus. Collecting the data helps to recognize the pattern, figuring out the predictability of the pattern gives a rate of change, and the finished calculated area reflects the accumulation. There you have it, three elementary- and middle-school connections to calculus are through patterns, rate of change, and area! But wait (as they say), there's more.

Calculus study brings in a whole new concept of ideas with a new set of symbols and vocabulary to describe these infinitely small changes (that's where limit comes in), changes so small that the difference does not matter much anymore for actual measurement. It can feel like first grade all over again, only older, with new concepts and new symbols. Calculus and precalculus students learn how limits in different situations, like increasing or decreasing inventory amounts, make use of these very small increments of change.

Archimedes used this notion of limit when he closed in on a number for pi with his ninety-six polygon sides with those internal and external polygons. Transitioning a bumpy rectangle area into a smoother circumference to calculate a circle's area also uses that notion of limits as the wedges get thinner and thinner. We can leave the details of those calculus spelunking explorations to the Math Aficionados as they learn all about those new concepts with different symbols.

Although the graph example in Figure 12.7 shows how changes happen with a line, the idea applies to curves as well, but with different trends and different triangle and rectangle sizes. The idea now is to show how all of those concepts that your child is learning in elementary, middle, and high school fit into that revered, and sometimes feared, pinnacle of math studies called calculus. Understanding the connections among these topics and knowing that all of these topics connect together can take some of the sting out of calculus fears.

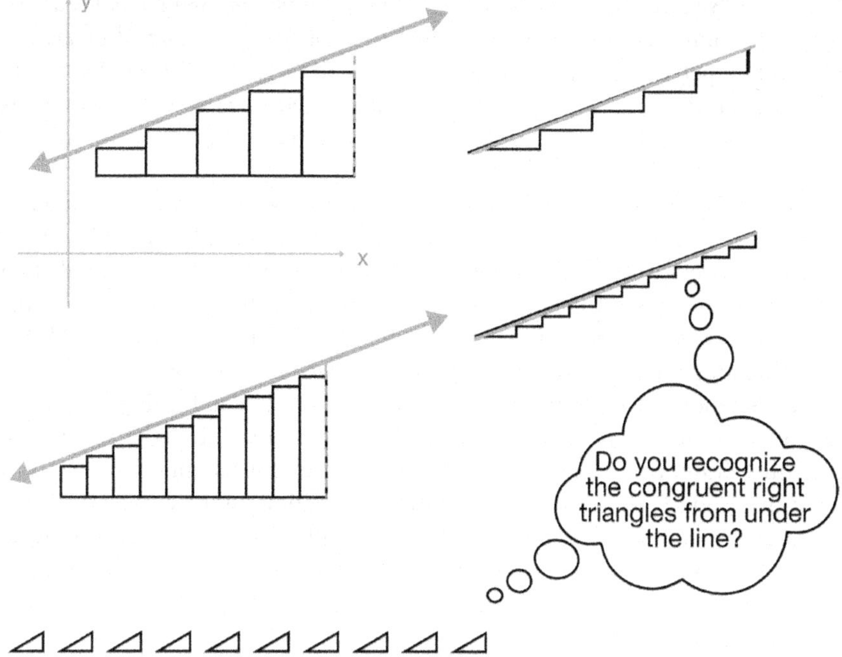

Figure 12.7   Areas, rate of change with slope, and congruent triangles come together in calculus

The linear trend graph example in Figure 12.7 shows a line placed on coordinate graph in a semitransparent background just for reference, not a specific linear equation. The graph also shows many rectangles topped with matching triangles underneath the line so that the changes (shrinking widths of the rectangles) are easier to see. Those rectangles underneath the line are reminders of the earlier predictable changes in rectangle patterns (Figure 12.3) when only one pair of sides changed. As the rectangular bars get thinner (changing the x measure using symbol Δx), the bases and heights of the triangles *on top of the rectangles under the line* get smaller.

The widths of these rectangles and triangles *change at the same rate* because they are part of the same measure underneath the line. As the widths of the rectangles get smaller, more rectangles appear—a lot more. The heights (or altitudes or other leg of the right triangle) are also changing. Yet all of the right triangles under this line, each time the widths shrink, stay congruent. How small will those triangles get on the top of each bar? Very, *very* small. In fact, the triangles will get so small and the rectangles will get so thin that the width differences eventually reach that previously mentioned limit.

The ratio describing *this change on the line border* goes by the name "slope" and is written as $\Delta y/\Delta x$ in that first-year algebra class. That same ratio between the Δy and the Δx, the change in the two legs of the right triangle, is also known as the tangent ratio in the right triangles. The tangent ratio is written as sine/cosine in a trig class and looks not-so-surprisingly very much like the slope ratio. The two expressions describe what's happening with the changes.

The concepts reviewed so far from your child's elementary math studies are pattern recognition, similar shapes, congruent shapes, right triangles, rate of change, areas, graphs, and upper-level algebra linear equations and trig ratios. Calculus becomes a compilation of all those topics, and more. Understanding is essential; shortcuts and sound bites will not help you or your child weave these ideas into a new world of measuring change.

The height of each rectangle-combination-with-triangle (trapezoid) underneath the line (or other curve) is dependent on the equation (also called a function) from earlier algebra classes. The idea is to keep the bases the same and draw the heights of the rectangle-triangle combination to reach the graph of the curve. The rectangle–triangle combination shows how the areas of those shapes shrink but can still be added together to find a total area under a line (or other graph) curve to show total accumulation, a total area between specific beginning and ending locations.

The slope of the line on the "hypotenuse side of the right triangle" can be thought of as getting closer and closer to the curve and eventually becoming the "slope of the curve at that point." The slope at that particular location on the curve also applies to a line tangent to the curve *at that same location*. Either way, the slope ratio for linear equations in algebra class (or other curves from algebra) and the tangent ratios in trigonometry class are connected to measuring changes in calculus and work together to describe the curve's behavior.

KEEP IN MIND

Many more examples for connecting figured algebra to written geometry are just waiting to be identified. Like Pappas, you can help your child see the mathematics in the shape relationships in their surroundings, filter out the surrounding irrelevant elements, and leave only the basic necessary measurements and lines in order to recognize the written geometry part of mathematics. Theon calculated square roots using figured algebra by drawing a diagram like the one in Figure 12.8, and mathematicians use this same diagram to show how this written geometry can look like multiplication with all kinds of numbers. Connections to the rescue!

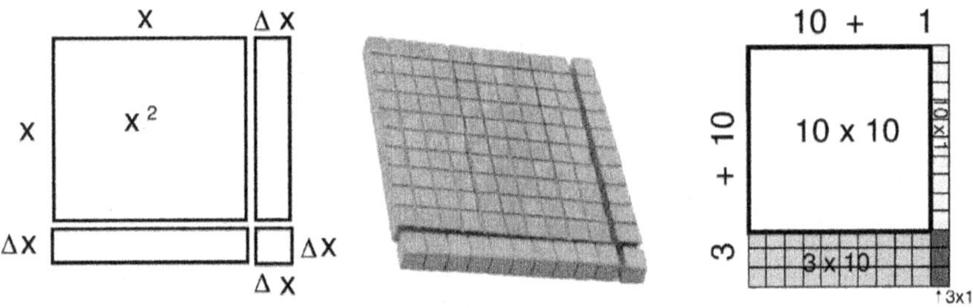

Figure 12.8 Theon's fourth-century diagram and the modern day area diagram have the same visual-spatial arrangement.

NOTES

1. Smith, *Agnesi to Zeno*, 121.
2. Draper, *Winning the Math Homework Challenge*, 124.
3. Sawyer, *Mathematician's Delight*, 183.
4. Tobias, *Overcoming Math Anxiety*, 232.
5. Strogatz, *Joy of X*, 132.

What Parents Can Do

The beginning of wisdom is "I do not know."[1]

—Gene Roddenberry, *Data*

Wisdom is that pot 'o gold reward for all that work to learn what you didn't know in the beginning, the kind of learning where imagination and curiosity are in the mix. Far too many parents believe that they are *supposed* to know not only the math, but know all about all the subjects their child is studying in school. Parents may be uncomfortable with Data's comment, even though admitting to not knowing something now is actually when real knowledge begins to gain a secure foothold for later.

Math tends to take the most punishment around Data's observation of "not knowing," sometimes because of parents' own past experiences, sometimes because a "new method" is now in vogue and threatening a parent's feelings of math equilibrium (whether tentative or confident), and sometimes because the instruction is missing the visual-spatial instruction mark. This book focuses on seeing math relationships through "visual-spatial eyes" to provide you and your child with the opportunity to "break through" and bridge negative past experiences.

THE BEST STRATEGY IS A TEAM STRATEGY

Practice saying, "I do not know," together with your child, and quickly tag on the word "yet" to the declaration! Get comfortable asking questions. Make learning a team effort and include visual-spatial strategies. Chances are that visual-spatial awareness and its application to mathematics is a new adventure for both you and your child. Visual-spatial learning is not just about pictures. Visual-spatial instruction involves a whole host of strategies that can capitalize on a child's visual-spatial strengths, including activities for classifying, sorting, storytelling (not just word problems), big overarching ideas, color-coding relationships, and hands-on materials.

These are only a few of the many ways to engage a child's creativity and imagination. Those project-driven learning experiences like Dixon's fourth-grade village in Chapter 1, Heather's eyeglasses for a great white whale in Chapter 12, and the ancient surveyors' tunnel model (also from Chapter 12), will challenge your child's curiosity to learn more about all sorts of cross-disciplinary information while they are blissfully unaware of *how much they are actually learning.*

Concept graphic organizers and mind maps provide a way to organize topics and ideas globally rather than through item-by-item or step-by-step tasks. Color-coding related pieces will emphasize the connections that highlight what would be helpful for children *to see on their own* (not limiting them to what we tell them about a relationship). Classifying and organizing (by what your child recognizes as characteristics) encourages him to see different perspectives and name them, and hands-on tactile projects help your child to piece things together so that the pieces fit in their own mind.

Invite your child to draw (or illustrate or diagram) their own versions of the concept or idea or skill because, as Sawyer suggests, "Even if the pictures are not good, the effort of making them will leave lasting traces in the mind and can cause the work to be remembered."[2] These strategies are the stuff of making mathematics (especially geometry) accessible to all children, especially to those who are visual-spatial learners.

Math Aficionados and mathematicians do not mind getting "wrong" answers or going down a "wrong-way math street" because those wrong answers expose errors and "misguided analogies and other woolly thinking, and bring the crux of the problem into sharper relief."[3] Mathematicians, Math Aficionados, and your child too can learn by asking questions in a safe judgment-free zone. Their questions help them identify the missing pieces or shake off "wooly thinking." Good questions don't have obvious answers. Productive questions pique curiosity and math wonderment. They allow your child to *learn new directions from* "wrong way" answers.

A VISUAL-SPATIAL PARENT-AND-CHILD STORY

Fran, a parent of a visual learner, described his own difficulties with math throughout school and did not want his visual-learning child to have the same experience. Throughout his elementary and high school years, he was repeatedly reminded that he was not "good at math" and suffered ridicule from his peers because he couldn't "keep up with the others." After he graduated high school and entered college, he hoped that "things would be different in math." He had failed the algebra course so many times that he was "almost" convinced that passing that algebra course was impossible.

Yet, he persisted because he *really didn't believe that success in math was out of his reach.* He did not give up. He eventually passed the necessary algebra courses and earned two degrees, one in Liberal Arts and the other in Science. His elementary-school testing for his Individualized Education Plan (IEP) included the Wechsler Intelligence Tests for Children (WISC). His results showed a Block Design score of 10, a Picture Arrangement score of 12, and an Object Assembly score of 12 (any score above 10 is considered a strength). These subtests are indicators of spatial abilities as determined by Beckman, Bannatyne, Kaufman, and Silverman.

No one suspected that Fran was a visual learner so his IEP extra math help consisted of auditory-sequential kinds of help—drill and practice, using even smaller steps in the step-by-step instruction—with the predictably unsuccessful results because these strategies don't work for visual learners. Fast forward to Fran's first-grade child who was beginning to voice her self-doubts about her ability in math even though at home she was designing and creating machines *with moving parts* using cardboard, sticks, and duct tape.

Fran recognized these feelings and was determined that his child was not going to suffer as he did. Times had changed, educational research had altered philosophies, and his own life experiences had made him aware of visual-spatial strategies. His daughter's math self-esteem began to turn around (and still continues to grow) *after* she started seeing math relationships by arranging blocks, drawing more pictures from her imagination to show her thinking (no problems here), and continuing her effort to explain (still having some difficulty but improving) as she prepares to enter fourth grade.

SUPPORT YOUR CHILD'S STRENGTHS

"It is impossible to be a mathematician without being a poet in soul" is a quote by Sofia Kovalevskaya.[4] Like Fran, Sofia Kovalevskaya's parents, and Sophie Germain's parents, your child needs your support to learn about mathematical relationships, especially if your child is a visual-spatial learner. Spatial ability is already hardwired in the brain at some level, so while your child may not grow up to be a poet or a mathematician, they do need to be encouraged to develop and use their spatial gifts in order to keep their openness to soul when they begins their math-learning journey.

Eighteenth-century Sophie Germain went against her native French culture's grain—and her family's wishes—to study mathematics. Eventually, her parents allowed her to follow her "unconventional" desire to study mathematics. Her contributions were stunning, but in order to communicate with other mathematicians, Germain was required to use a male pen name, Auguste Antoine Le Blanc. Despite these odds (and other cultural roadblocks), she was awarded a prize for her *Memoir on the Vibrations of Elastic Plates*. Talk about spatial abilities and supportive parent encouragement!

Kovalevskaya's nineteenth-century parents also went against the prevailing gender-biased European culture and supported her "soul desire" to learn mathematics. Eventually, she became the first woman since the Renaissance to receive a doctorate in math. Single women of any age could not travel alone during the nineteenth century without their parents' permission, so her parents did what was necessary to make it possible for her to travel freely and learn from the best mathematicians of the time. Kovalevskaya earned the Academy of Science's *Prix Bordin* for authorship of a paper "On the Problem of the Rotation of a Single Body About a Fixed Point."

The experiences of the past can help us learn about what can be accomplished in the future. Eighteenth- and nineteenth-century gender prejudices may supposedly be behind us, but parental support and encouragement is still very much needed in the present and the future of our children, both boys and girls. Mathematics is not a special gift reserved for the elite, or limited by gender, culture, or age. Mathematics is an equal opportunity system just waiting to be explored, using any and all of an individual's learning strengths.

Help your child build models with manufactured toys or cardboard tubes, duct tape, and creativity. Not all tools for learning mathematics are on the retail shelf or look like online practice sheets. Many of the tools for learning about math relationships are in kitchen drawers, recycle bins, storage closets, or garages. The most essential ingredients are imagination, creativity, and curiosity. The end goals for any math learning activity are learning how to assemble (composition) and disassemble (decomposition), see structure and symmetry, and predict what something would look like from all directions.

The written part of the geometry (the algebra and the arithmetic) must eventually support the solutions, but the spatial exploration and experimentation should come first so that the written symbols and explanations will make sense. Remember, a parent's job is to keep those mathematical "soul fires kindled," like Sophie Germain's parents and Sofia Kovalevskaya's parents did. It is indeed easier to help your child if you understand the ideas, even if you don't know the solution—yet. "The silver lining is that even wrong answers can be educational . . . as long as you realize they are wrong."[5]

FOUR POSSIBLE MATH PROFILES TO CONSIDER

Children learn how to cope with math failure. If they can't "do the math" quickly, they make up something—the faster, the better. What does it mean to "do math" and what on earth does speed have to do with it when it has taken centuries for brilliant mathematicians to understand and organize some of these relationships? What we now call simple addition or multiplication "math facts" took five centuries to develop. And without Aryabhata's fifth-century place value, the phrases "two plus eight equals ten," "three times four equals twelve," or Huck Finn's "six times seven is thirty-five . . . even if I was to live forever"[6] have no meaning.

The four student profiles in Figure C.1 represent students everywhere, with different names and different ages. Some of the behaviors could represent visual-spatial learner survival strategies in a traditional verbal-sequential math classroom. Some of these behaviors—not all are math-related—represent emotional failures and intimidations. Because these intimidating and failing events happened in association with a math situation, the connection is indelibly etched into the psyche around math. What kind of math student are/were you? Were you a visual learner in an auditory class? Do you recognize any of these behaviors in your child?

These four profiles show some of the characteristic behaviors that represent children's coping and reasoning strategies when their learning experiences do not correlate with their strengths. Rote learners turn to memorization out of desperation, Perplexed learners learn that what they think doesn't matter, and the Gambler needs to get it over with, the faster the better. The visual learning "cloud" overlaps the Perplexed, Rote, and Gambler profiles, and embraces the Puzzler who was fortunate enough to have identified their learning strengths. These profiles are here to make you think—about your own past experiences and your child's experiences now.

There is no magic pill or perfect learning environment or guaranteed method that works successfully for all children. Learning is personal, whether or not someone is

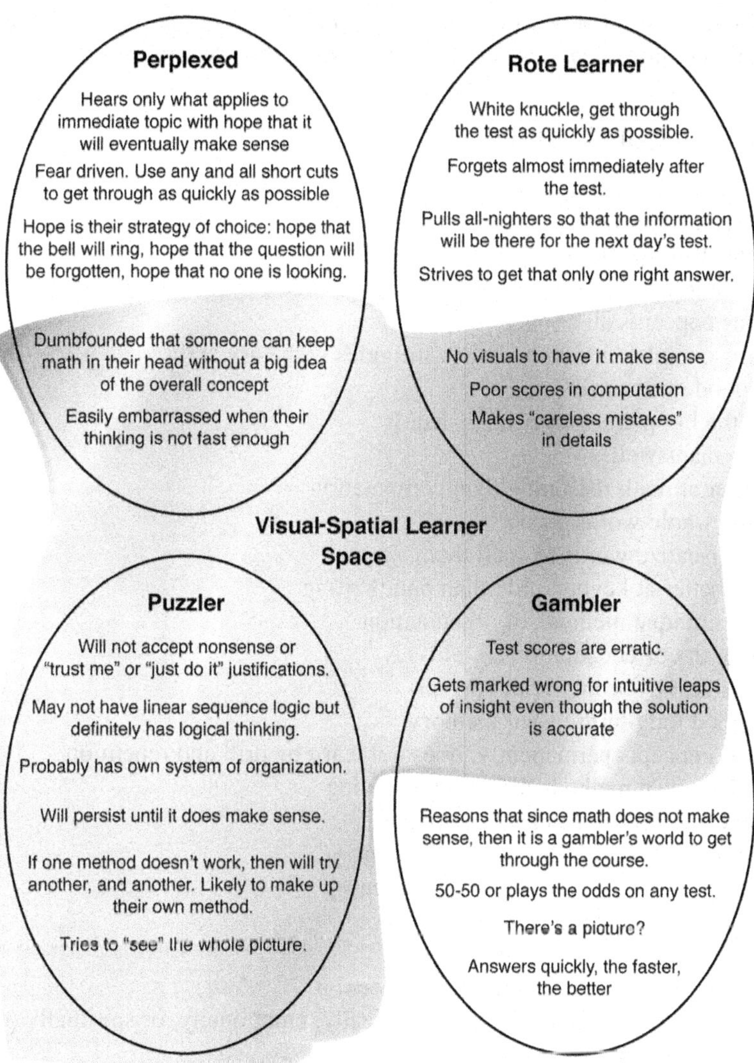

Figure C.1 The gray cloud space overlap shows how visual spatial learnerscan be reflected in the four imaginary student profile examples.

a visual learner. Children need to be involved in their learning so that they have the chance to take some control over their learning. The best way for them to be involved is for them to have a "stake in the process" through experiments and explorations that pique their curiosity. The more they participate, the better chance that they have to have more glorious weeks of learning throughout elementary, high school, and college.

Information about visual-spatial learning and lists of preferred strategies are in every library and on many websites. A comprehensive list was developed through Linda Silverman's research and is listed below.[7] As Silverman's research indicates, it is futile to expect success from visual-spatial learners using only auditory-sequential teaching methods. Your child does not have to have all of the spatial characteristics to be frustrated, just enough to interfere in his successful learning.

The Visual-Spatial Learner

- Thinks primarily in images
- Has visual strengths
- Relates well to space
- Is a whole-part learner
- Learns concepts all at once
- Learns complex concepts easily; struggles with easy skills
- Is a good synthesizer
- Sees the big picture; may miss details
- Reads maps well
- Is better at math reasoning than computation
- Learns whole words easily
- Must visualize words to spell them
- Much better at keyboarding than handwriting
- Creates unique methods of organization
- Arrives at correct solutions intuitively
- Learns best by seeing relationships
- Has good long-term visual memory
- Learns concepts permanently; does not learn by drill and repetition
- Develops own methods of problem solving
- Is very sensitive to teachers' attitudes
- Generates unusual solutions to problems
- Develops quite asynchronously (unevenly)
- May have very uneven grades
- Enjoys geometry and physics
- Masters other languages through immersion
- Is creatively, mechanically, technologically, emotionally, or spiritually gifted
- Is a late bloomer

The Auditory-Sequential Learner

- Thinks primarily in words
- Has auditory strengths
- Relates well to time
- Is a step-by-step learner
- Learns by trial and error
- Progresses sequentially from easy to difficult material
- Is an analytical thinker
- Attends well to details
- Follows oral directions well
- Does well at arithmetic

- Learns phonics easily
- Can sound out spelling words
- Can write quickly and neatly
- Is well organized
- Can show steps of work easily
- Excels at rote memorization
- Has good auditory short-term memory
- May need some repetition to reinforce learning
- Learns well from instructions
- Learns in spite of emotional reactions
- Is comfortable with one right answer
- Develops fairly evenly
- Usually maintains high grades
- Enjoys algebra and chemistry
- Masters other languages in classes
- Is academically talented
- Is an early bloomer

KEEP IN MIND

Whether your child is a visual-spatial learner or prefers the step-by-step sequence of the auditory-sequential learner, nurturing your child's imagination at every age and at every level of mathematics will always pay off multifold with positive self-esteem and math-content successes on many levels. Your child's path to the success kingdom is through connections, so help them look for connections at every opportunity. Even if those connections are not accurate at first, they will help you and your child uncover wooly thinking. You can't go wrong if you apply these four basic guidelines:

- Let them correct their own errors.
- Provide a safe place for them to express mathematics *the way that they see it.*
- Provide a clue about when to stop nonproductive efforts (but remember that this does not mean to make your child do the math *your* way instead of allowing him to devise their way to make it make sense to them *and be accurate*).
- Show, whenever possible, what the desired end looks like, and let them fill in a process.

As many others have said about making positive changes, "If not now, when? If not me, who?" Yes, *you* are the "who" for your child and the time is now! Spatial-sense, number-sense, insights, jigsaw thinking, flexibility, imagination, organization, connections, relationships, patterns—all of these, and more from the other books in this series, will provide you and your child with the necessary tools to succeed in mathematics. Mathematics isn't just about algebra, logic, or juggling symbols; it is also about visualization, imagination, and making sense by whatever means necessary for it to make sense to you.

One may pay lip service to Descartes and Grothendieck when they wish that geometry be reduced to algebra, or to Russell and Gentzen when they command that mathematics become logic, but we know that some mathematicians are more endowed with the talent of drawing pictures, others with that of juggling symbols, and yet others with the ability of picking the flaw in the argument.[8]

—Gian-Carlo Rota

NOTES

1. Video clip from "Star Trek, The Next Generation," Season 2, Episode 2, https://www.youtube.com/watch?v=8eDYVtPwWiM.
2. Sawyer, *Vision in Elementary Mathematics*, 12.
3. Strogatz, *Joy of X*, 63.
4. Smith, *Agnesi to Zeno*, 155.
5. Strogatz, *Joy of X*, 63.
6. Twain, *The Adventures of Huckleberry Finn*, 33.
7. https://www.time4learning.com/visual-spatial-learners.shtml
8. Gian-Carlo Rota quoted in Davis, *The Mathematical Experience*, xviii.

Glossary

This glossary is generated by definitions provided by students. Please refer to a mathematical dictionary for formal mathematical definitions.

absolute value – distance between two locations without positive or negative notation

acute – an angle "adjective" that describes an angle that is less than 90°

additive inverse – zero, the "subtraction" to get to zero

adjacent (angle or side) – adjective that describes how two sides or two angles are arranged; next to each other

algorithm – operation or procedure with numbers; add, subtract, multiply, divide, and all the other defined procedures

alternate (angles) – adjective that describes how two angles are arranged; across a transversal

altitude (height) – a one-dimensional line that defines the distance in a two-dimensional polyhon from one side to its opposite vertex or side

area – coverage inside of a shape

auditory-sequential – learns best by hearing step-by-step directions

average – when everything in the number list is added together and then that total is divided by the number of things in the list

circumference – the length around a circle space

circumscribed – depends on which one is being circumscribed by which shape; "being circumscribed by" is a phrase that indicates which shape is doing the circumscribing and is on the "outside"

commutative – the name given to the situation when the operations for addition and multiplication work both ways

complementary (angles) – an angle "adjective" that describes how two angles will add to 90°

composition – building a term or shape with component pieces

congruent – the name assigned to shapes when their number measures are equal

coordinate points – a pair of numbers that shows where two number directions meet

cosecant – the reciprocal ratio of sine; the ratio is hypotenuse/opposite.

cosine – the shift of a sine curve; the ratio of adjacent side to hypotenuse side in a right triangle

cotangent – the reciprocal ratio of tangent; the ratio is cosine/sine.

cylinder – a 3-D tube-looking shape

decomposition – taking apart a term or a shape into component parts

diagonal – the one-dimensional line that connects two corners in a polygon

diameter – the one-dimensional line that connects the opposite ends on a circle circumference through the center

directional reading – how your eyes track while reading something

e – 2.718266254. . . irrational and transcendental number

eccentricity – a ratio for conic section curves that describes closeness to a circle graph

ellipse – round shape that has a major and minor axis; looks more like an oval

equilateral – all sides are congruent; refers to a regular triangle that is also equiangular

exponents – the small number in the upper right side of a term that tells how many times that term is multiplied by itself

factor – decomposing numbers by using division; no fractions

fractions – part-of-it as compared to all-of-it ratio

imaginary number – a number that does not belong to the Real number set

inscribed – when a polygon is inscribed then it is "inside" of a circle and the vertices of the polygon touch the circle's circumference

integer – a whole number that has either a + or a – indicator; no fractions allowed

inverse operation – the opposite operation that takes numbers back to their "origin"; for addition, the origin is 0; for multiplication, the "origin" is 1

isosceles – the adjective assigned to a triangle with two congruent sides, or quadrilateral when the two opposite sides are congruent

logarithm – how to write a number that equals the exponent

obtuse – an angle "adjective" that describes an angle that is greater than 90° and less than 180°

parabola – a curve that is symmetric and has an equation with the highest exponent of 2, like $y=x^2$

parallel – two lines or line segments that will never touch

parallelogram – a quadrilateral with opposite parallel sides

perimeter – the distance around the boundary of a shape

perpendicular – when a vertical line crosses a horizontal line; it makes a 90° angle

phi – 1.618 . . . irrational and transcendental number

pi – 3.14159 . . . irrational and transcendental number

polygon – straight sided, two-dimensional shapes named by the number of sides that they have (triangle 3, quadrilateral 4, pentagon 5, hexagon 6, and so on)

polyhedra – straight-sided, three-dimensional shapes

prism – many-sided polyhedra

Pythagorean theorem – $a^2 + b^2 = c^2$ when "a" and "b" are legs and "c" is the hypotenuse in a right triangle

quadrilateral – four-sided polygon that has several subnames: kite, trapezoid, parallelogram, rectangle, rhombus, square

radian – a straight radius length that has been curved to fit on the circumference of the same circle; approximately between 56° and 57°

radius – half of a diameter

reflection – a transformation that reflects a shape across a line of symmetry

regular – all sides of a shape have equal measures and all angles are equal to each other

rhombus – a quadrilateral with all adjacent sides congruent but not necessarily a square

right (angle) – an angle "adjective" that describes an angle that is equal to 90°

rotation – a transformation that rotates a shape

scalene – a side length "adjective" that describes when none of the sides of a triangle are congruent

secant – reciprocal ratio to cosine ratio and is written as hypotenuse/adjacent

similarity – same shape with proportional sides

sine – a continuous curve in trig; the curve goes through the origin (0, 0) and is the ratio between the opposite side and the hypotenuse in a right triangle

slide – a transformation that slides a shape along a line

slope (of a line) – the ratio used as a coefficient to x in a linear equation; the ratio tells how high the line rises on y in relation to the run on the x axis between two points on the same line

square numbers – when a number of items can be arranged into a square

supplementary (angles) – an angle "adjective" that describes that describes how two angles will add to 180°

symmetry – when two sides of a shape exactly match

tangent – a noncontinuous curve in trig; the curve goes through the origin (0, 0) and is defined by the ratio of the sine over the cosine ratios of the sides of a right triangle

transformation – description for a shape orientation change; flip, slide, turn or reflection, slide, rotation

transversal – a line that crosses over two parallel lines

trapezoid – a quadrilateral that has exactly two parallel sides

triangle – three-sided polygon with different "adjectives" that describe angles (acute, obtuse, right) and sides (scalene, equilateral, isosceles)

trigonometry – a kind of math that studies the ratios of sides in a right triangle that match to angles

vertex – a corner of a polygon or polyhedra

volume – what fills up the contents of a 3-D shape

References

Abbott, Edwin Abbott. *Flatland*. New York: Dover Publications, 1952.

Apleman, Maja, and Julie King. *Exploring Everyday Math*. Portsmouth, NH: Heinemann Publications, 1993.

Armstrong, Thomas. *Multiple Intelligences in the Classroom*. Alexandria, VA: Association for Supervision and Curriculum Development, 1994.

Baretta-Lorton, Mary. *Workjobs for Parents, Activity-Centered Learning in the Home*. Menlo Park, CA: Addison-Wesley, 1975.

Bennett, Dan. *Pythagoras Plugged In*. Emeryville, CA: Key Curriculum Press, 2003.

Boyer, Carl B. *A History of Mathematics*. Hoboken, NJ: Wiley, 1968.

Brosterman, Norman. *Inventing Kindergarten*. New York: Harry N. Abrams, 1997.

Carroll, Lewis (Charles Dodgson). *Alice's Adventures in Wonderland*. New York: Macmillan, 1865.

Carter, Beth W. "Move Over Frank Lloyd Wright." *The Arithmetic Teacher* 33 (September 1985): 8–11.

Common Core State Standards for Mathematics, Common Core State Standards Initiative, 2014, www.corestandards.org.

Covey, Stephen. *The Seven Habits of Highly Successful People*. New York: Simon and Shuster, 1990.

Davis, Philip J., and Reuben Hersh. *The Mathematical Experience*. Boston: Houghton-Mifflin, 1981.

Dixon, John Philo, PhD. *The Spatial Child*. Springfield, IL: Charles C. Thomas, 1983.

Doczi, György. *The Power of Limits*. Boulder, CO: Shambala Publications, 1981.

Draper, Catheryne. *The Algebra Game: Linear Graphs, Quadratic Equations, Conic Sections, and Trig Functions*. Rowley, MA: Didax Education Inc., 2016.

———. *How the Math Gets Done: Why Parents Don't Need to Worry About New vs. Old Math*. Lanham, MD: Rowman and Littlefield, 2017.

———. *Solving with Pythagoras*. Salem, MA: The Math Studio, Inc., 2005.

———. *User-Friendly Math for Parents: Learning and Understanding the Language of Numbers Is Key*. Lanham, MD: Rowman and Littlefield, 2017.

———. *Winning the Math Homework Challenge: Insight for Parents to See Math Differently*. Lanham, MD: Rowman and Littlefield, 2017.

Eastaway, Rob, and Mike Askew. *Old Dogs, New Math: Homework Help for Puzzled Parents*. New York: The Experiment Publishing, LLC, 2010.

Enzensberger, Hans Magnus. *The Number Devil: A Mathematical Adventure*. New York: Metropolitan Books, 1997.

Eves, Howard. *An Introduction to the History of Mathematics.* New York: Holt, Rinehart, and Winston, 1969.

Ghyka, Matila. *The Geometry of Art and Life.* New York: Dover Publications, 1977.

Grandin, Temple. *Thinking in Pictures.* New York: Vintage Books, 2006. http://www.grandin.com/inc/visual.thinking.html

Halmos, Paul. "Mathematics as a Creative Art." The Royal Society of Edinburgh Year Book 1973 (Session 1971–1972). http://www-history.mcs.st-andrews.ak.ul/Extras/Creative_art.html

Heath, T. L. *A History of Greek Mathematics. Vol. 1: From Thales to Euclid.* New York: Cambridge University Press, 2014.

Hersh, Reuben, and Vera John-Steiner. *Loving + Hating Mathematics.* Princeton, NJ: Princeton University Press, 2011.

Hilton, Peter, and Jean Pedersen. *Fear No More: An Adult Approach to Mathematics.* Menlo Park, CA: Addison Wesley Publishing Company, 1983.

Hogben, Lancelot. *Mathematics for the Millions.* New York: Norton, 1951.

Jacobs, Harold R. *Mathematics: A Human Endeavor.* New York: W. H. Freeman and Company, 1982.

Johnson, Art. *Building Geometry.* Menlo Park, CA: Dale Seymour Publications, 1997.

Joyce, David E. "Euclid's Elements." http://aleph0.clarku.edu/~djoyce/java/elements/Euclid.html

Kasner, Edward, and James Newman. *Mathematics and the Imagination.* New York: Simon and Schuster, 1940.

Kenschaft, Patricia Clark. *Math Power: How to Help Your Child Love Math, Even If You Don't.* New York: Addison Wesley Longman,1997.

King, Julie. "The Power to Make Predictions." *Connects Magazine* (January, February 2007): 8–10.

Kolpas, Sidney J. *The Pythagorean Theorem, Eight Classic Proofs.* Palo Alto, CA: Dale Seymour Publications, 1992.

Maor, Eli. *e: The Story of a Number.* Princeton, NJ: Princeton University Press, 1994.

McNeil, Ruth. "A Reflection on When I Loved Math and Why I Stopped." *Journal of Mathematical Behavior,* 1988.

NCTM. *Principles and Standards for School Mathematics.* Reston, VA: National Council of Teachers of Mathematics, 2000.

Newman, Rochelle, and Martha Boles. *Universal Patterns.* Bradford, MA: Pythagorean Press, 1992.

Pólya, George. *How to Solve It.* Second edition. Garden City, NY: Doubleday Anchor Books, 1957.

———. *Mathematical Discovery.* Vol. 1. New York: John Wiley and Sons, 1967.

———. *Mathematical Discovery.* Vol. 2. New York: John Wiley and Sons, 1968.

———. *Mathematics and Plausible Reasoning.* Vol. 1. Princeton, NJ: Princeton University Press, 1954.

Roy, Arijit. *The Enigma of Creation and Destruction.* Bloomington, IN: AuthorHouse, 2011.

Sawyer, W. W. *Mathematician's Delight.* Baltimore, MD: Pelican Books, 1943.

———. *A Prelude to Mathematics.* Baltimore, MD: Pelican Books, 1955.

———. *Vision in Elementary Mathematics.* Baltimore, MD: Pelican Books, 1964. pp. 298–99.

Silverman, Linda Kreger, PhD, and Buck Jones. *Upside Down Brilliance—Strategies for Teaching Visual-Spatial Learners.* Denver, CO: DeLeon Publishers, 2002.

Slatner, David. *The Joy of π.* New York: Walker and Company, 1997.

Smith, Sanderson M. *Agnesi to Zeno: Over 100 Vignettes from the History of Math.* Emeryville, CA: Key Curriculum Press, 1996.

Sousa, Dr. David A. *How the Brain Learns Mathematics.* Thousand Oaks, CA: Corwin Press, 2000.

Stewart, Ian. *Nature's Numbers, The Unreal Reality of Mathematics*. New York: BasicBooks, 1995.

———. *Taming the Infinite: The Story of Mathematics from the First Numbers to Chaos Theory*. London: Quercus Publishing, 2008.

Strogatz, Steven. *The Joy of X*. Boston: Houghton Mifflin Harcourt, 2012.

Su, Francis E., et. al. "Eccentricity of Conics." Math Fun Facts, 1999–2010. http://www.math.hmc.edu/funfacts

Tobias, Sheila. *Overcoming Math Anxiety*. Boston: Houghton Mifflin Company, 1978.

Twain, Mark (Samuel L. Clemens). *The Adventures of Huckleberry Finn*. New York: Harper and Brothers, 1884.

Wells, David. *Curious and Interesting Numbers*. Hamondworth, UK: Penguin Books, 1986.

Wertheim, Margaret. *Pythagoras' Trousers*. New York: Norton, 1997.

West, Thomas G. *In The Mind's Eye*. Amherst, MA: Prometheus Press, 2009.

Wheatley, Grayson H. "Image Maker, Developing Spatial Sense." *The Arithmetic Teacher* 18 (February 1999): 374–78.

———. "Spatial Sense and Mathematics Learning." *The Arithmetic Teacher* 37 (February 1990): 10–11.

Williams, G. Arnell. *How Math Works*. Lanham, MD: Roman and Littlefield Publishers, 2013.

MATERIALS

Algebra Game: Linear graphs, quadratic equations, conic sections, and trig functions
Algebra tiles
Anglegs
Base ten blocks
Creativity
Chips, disks, color counters
Exploragons
Geometry sketchpad
Graphic organizers
Grid paper
Imagination
Legos
Linking cubes
Markers—colors
Magformers
One-inch blank or color wooden cubes
One centimeter blank or color cubes
Pantograph
Positive and negative rotating number line
Polydrons
Patty paper (tracing paper)
Toothpicks, straws, chopsticks (building shapes), and putty to build 2D and 3D shapes
Unifix cubes
Visualization
Weights for plumb line perpendicular
Yarn

About the Author

Catheryne Draper's tenure in math education has crossed the half-century mark. After graduating with a BS in Mathematics and an MEd degree in Mathematics Education and Supervision from the University of Georgia, she has been a high-school teacher before serving as a district supervisor, state-level advisor, professional-development consultant, and math coach/teacher for both large urban and smaller public and private schools. She has taught math at all levels, from kindergarten through college, as well as maintaining for almost thirty-five years an ongoing practice working with individual children and adults who "see" math differently.

She has worked as a math editor for textbook and supplementary materials in regular education and special education. She opened The Math Studio in the early 1980s to pursue her passion of providing visual and tactile instruction for all students. Her twenty-five-plus year tenure working with clients of the Massachusetts Rehabilitation Commission brought predictably successful results as she helped them succeed in their math learning goals. As part of her work at The Math Studio, she developed and published *The Algebra Game Program*, an instructional and assessment card program for classrooms and individuals.

In addition to teaching and working in educational publishing, Draper has written articles for and contributed to many journal publications about mathematics education pedagogy. Her previously published books include *Winning the Math Homework Challenge: Insights for Parents to See Math Differently* and *User-Friendly Math for Parents: Learning and Understanding the Language of Numbers Is Key*, and *How the Math Gets Done: Why Parents Don't Need to Worry About New vs. Old Math.*

While in Georgia, Draper was instrumental in developing a K–8 instructional and assessment tool with a computer-driven assessment component and was responsible for bringing computer terminals into the district's high-school curriculum, predating today's technology involvement by decades. After leaving Georgia, Draper worked with national assessment companies and publishing companies prior to opening the doors of The Math Studio to further influence positive changes in mathematics education.